Analysis of the K-Epsilon
TURBULENCE MODEL

OTHER TITLES IN THE SAME SERIES

GLOBAL CLASSICAL SOLUTIONS FOR QUASILINEAR HYPERBOLIC SYSTEMS, by LI TA-TSIEN. 1994, 328 pages.

RECENT ADVANCES IN PARTIAL DIFFERENTIAL EQUATIONS, by M.-A. HERRERO and E. ZUAZUA. 1994, 160 pages.

RAM Research in Applied Mathematics
Series Editors : P.G. CIARLET and J.-L. LIONS

Analysis of the K-Epsilon TURBULENCE MODEL

B. MOHAMMADI
Research Engineer at INRIA, Le Chesnay

O. PIRONNEAU
Professor at the University Pierre et Marie Curie, Paris VI

JOHN WILEY & SONS
Chichester • New York • Brisbane • Toronto • Singapore

1994

MASSON
Paris • Milan • Barcelona

Figures on the cover represent five turbulent flows simulations computed by P. Woodward, B. Mohammadi, M. Ravachol, M. Mallet and M. Brachet.

La collection ***Recherches en Mathématiques Appliquées*** *a pour objectif de publier dans un délai très rapide des textes de haut niveau en Mathématiques Appliquées, notamment :*

— des cours de troisième cycle,
— des séries de conférences sur un sujet donné,
— des comptes rendus de séminaires, congrès,
— des versions préliminaires d'ouvrages plus élaborés,
— des thèses, en partie ou en totalité.

Les manuscrits, qui doivent comprendre de 120 à 250 pages, seront reproduits directement par un procédé photographique. Ils devront être réalisés avec le plus grand soin, en observant les normes de présentation précisées par l'Éditeur.
Les manuscrits seront rédigés en français ou en anglais. Dans tous les cas, ils seront examinés par au moins un rapporteur. Ils seront soumis directement soit au

The aim of the ***Recherches en Mathématiques Appliquées*** *series (Research in Applied Mathematics) is to publish high level texts in Applied Mathematics very rapidly :*

— Post-graduate courses
— Lectures on particular topics
— Proceedings of congresses
— Preliminary versions of more complete works
— Theses (partially or as a whole)

Manuscripts which should contain between 120 or 250 pages will be printed directly by a photographic process. They have to be prepared carefully according to standards defined by the publisher.
Manuscripts may be written in English or in French and will be examined by at least one referee.
All manuscripts should be submitted to

Professor P.G. Ciarlet, Analyse numérique, T. 55,
Université Pierre et Marie Curie, 4, place Jussieu, 75005 Paris
soit au / or to
Professor J.-L. Lions, Collège de France,
11, place Marcelin-Berthelot, 75005 Paris

A catalogue record for this book is available from the British Library.

ISBN Masson : 2-225-84391-0 **ISBN Wiley** : 0-471-94460-2

ISSN : 0298-3168

MASSON S.A. 120, bd Saint-Germain, 75280 Paris Cedex 06
JOHN WILEY AND SONS Ltd Baffins Lane, Chichester, West Sussex PO 19 1UD, England

If God is not in mathematics, then where is He? [1]
M.L. Parashar

FOREWORD

Part of the material of this book was taught at the graduate level at the University of Houston in 1991 and at the University of Paris in 1992 and 1993. It is intended to applied mathematicians interested in numerical simulation of turbulent flows. The book is centered around the $k - \epsilon$ model but it deals also with other models such as subgrid scale models, one equation models and Reynolds Stress models.
The reader is expected to have some knowledge of numerical methods for fluids and if possible some understanding of fluid mechanics, its partial differential equations and their variational formulations.

In this book we have tried to present the $k - \epsilon$ model for turbulence in a language familiar to applied mathematicians and hopefully stripped bare of all the technicalities of turbulence theory. We attempt to justify the model from a mathematical stand point rather than from a physical one. We investigate the numerical algorithms, and present some theoretical and numerical results.

The reader should not lose sight of the fact that Turbulence Modeling is still very much in its infancy; the $k - \epsilon$ model is one possibility, much in fashion at the time of writing, but far from being the miraculous tool.

The material presented here is the result of ten years of research by many of our close collaborators. In particular we would like to thank

1 तत् त्वम् असि ("Tat Twam Asi", Isha Upanishad).

C. Begue, B. Cardot, T. Chacon, D. Franco, F. Ortegon and C. Parès who's doctoral research have contributed to the understanding of the $k - \varepsilon$ model in the light of multiple scales expansion and frame invariance; are included also in this book very significant contributions done by our close colleagues, C. Bardos, R. Lewandowski, D. Maclaughin and G. Papanicolaou. Our views have been strongly influenced by the determining work of D. Vandromme, and P.L. Viollet, and our thanks go also to J.P. Chabart, F. Coron, A. Dervieux, D. Laurence, P. Le Tallec, J. Periaux, P. Perrier, Ph. Rostand for their collaboration. Several numerical results were kindly provided by M. Brachet, F. Hecht, M. Guillard, M. Mallet, M. Ravachol, and P. Woodward. Finally we are grateful to Ph. Ciarlet and J.L. Lions for their stimulating support.

CONTENTS

NOTATIONS

References

References such as Kolmogorov[1941] refer to publications listed at the end of the book by alphabetic order and by the year of publication. Reference to a paragraph in the book uses §. Thus §3.2 refers to Paragraph 3.2 in the same chapter while §IV.3.2 refers to Paragraph 3.2 in Chapter 4.

Geometrical quantities

Ω open set of R^d, $d = 2$ or 3
x point in Ω, t time in $]0, T'[$ (T is the temperature).
Γ boundary of Ω
n, s, τ normal and tangents to Γ

Operators and Tensors

In the list below p, q are scalar valued functions, u, v are vector valued functions and A, B are 2nd order tensors (or $d \times d$ matrices).
$\partial_t u$ Partial derivative in t of u; $\partial_{tt} u$ is the 2nd derivative.
$\partial_j u$ Partial derivative of u with respect to x_j.
∇p Gradient of p (a vector of ith component $\partial_i p$).
∇u 2nd order tensor $(\nabla u)_{ij} = \partial_i u_j$.
$\nabla \cdot u$ Divergence of $u = \sum_i \partial_i u_i$.
$\nabla \cdot A$ Vector with j^{th} component $= \sum_i \partial_i A_{ij}$.
$\nabla \times u$ The curl of u.
$\nabla \times q$ in dimension 2 only: stands for $\nabla \times (0, 0, q)^T$.
$u \cdot v$ Scalar product.
$u \nabla u$ $= u_i \partial_i u$.
$A : B$ $= \sum_{ij} A_{ij} B_{ij}$.
$u \otimes v$ 2nd order tensor (or a $d \times d$ matrix), $(u \otimes v)_{ij} = u_i \otimes v_j$.
$\equiv$ Same as = but used when right sides define left sides.
$< u >$, $\overline{u}$ mean of u.
$Span\{u, v...\}$ is the vector space spanned by those vectors.

Fluid quantities

u fluid velocity, U mean velocity ($u = U + u'$)
ρ density
p pressure
T temperature
E total energy per unit volume
k turbulent kinetic energy
ϵ rate of dissipated turbulent energy

INTRODUCTION

A century has passed since Reynolds' pivoting work and yet fluid turbulence is still one of the greatest stumbling block for scientific and technological developments. From stars to rivulets, turbulence is almost impossible to predict. However even if turbulence is not understood in the sense of macroscopic modeling it seems that computers simulations of fluids will allow engineers to evaluate their design. So there is hope.

At the time of writing only a few simple configurations can be simulated directly with some accuracy. Too many points are necessary and so only the very simple geometrical configurations can be handled, such as homogeneous turbulence behind a grid, turbulence in a straight pipe or above a flat plate.

Thus engineers have proposed new sets of equations to describe the mean of a turbulent flow; the $k-\epsilon$ model is one such set. It was proposed by Launder and Spalding in the seventies[1972] and it is still very much used today even though more sophisticated models have been suggested (e.g. Reynolds stress models, Launder[1992]).

On the mathematical side, the situation is far from being understood. Several theories are being studied such as Chaos (Berger et al [1984]), Inertial Manifolds (Foias et al[1985]), multiple scale expansions and Homogenization (McLaughin et al [1982]), the theory of propagation of singularities (Serre[1991], Tartar[1990])... None of them provide the engineers with design tools in the presence of turbulence.

Thus in want of a better tool and for several more years to come, it seems that we will have to rely on ad-hoc models such as the $k-\epsilon$ model.

However the $k-\epsilon$ model is not so easy to use and so this is where this book fits in. Indeed it is not clear in the first place that the model is well posed or that it produces physically relevant results (a positive energy, for example). Furthermore it is numerically rather unstable or, at any rate rather hard to use. So we have gathered in this book the mathematical results available on the model and its numerical approximations.

The number of scientific articles on the $k - \epsilon$ model is phenomenal! A historical review would be too hard. If we can risk some bibliographical comments we may say that the idea of using two convection diffusion equations for the modeling of turbulent quantities was around in the sixties (see for example von-Karman[1948] and Rotta[1951]) but it is really in Launder-Spalding [1972] that the model is presented. Among the landmark papers in the field are: Patankar [1980], Rodi[1980], Viollet[1981]...For compressible fluids we have followed Vandromme[1983].

Since this book deals with mathematical and numerical tools for the $k-\epsilon$ model it begins by recalling the general equations for Newtonian fluids, the Navier-Stokes equations, and some of the mathematical results on existence and uniqueness and long time behavior. In *Chapter 1* a brief review of boundary layers is also given, along with an introduction to wall laws.

The book has three parts:

I*Incompressible flows.*

The basic concepts are presented on this simpler case.

II*Compressible flows.*

This part contains an extension of the previous tools to the full system of fluid mechanics

III*Convection of micro-structures.*

This part deals with a partial justification of the $k - \epsilon$ model by a multiple scale expansion and homogenization.

In *Chapter 2* the basic concepts of homogeneous turbulence are given together with a numerical method for its computer simulation. In particular Kolmogorov's 5/3 law is given here.

In *Chapter 3* we present the concept of turbulence modeling: the now classical analysis of Reynolds and the concept of filters and Reynolds stresses. We show also that frame invariance limits the free parameters of any model based on Reynolds' hypothesis to two unknown scalar functions only. The simplest subgrid scale turbulence model, due to Smagorinsky, is presented here as an illustration. This model gives to

us an opportunity to present two classical approximations of the Navier-Stokes equations by the Finite Element Method; a technique particularly well adapted to industrial configurations with complex boundaries.
In *Chapter 4* we introduce the $k - \epsilon$ model and recall the hypothesis on which it is based. We discuss the determination of the constants of the model and the boundary conditions. The low Reynolds number extension is also presented here.
Positivity, well posedness, numerical methods and performances of the model are discussed in *Chapter 5.*
Finally in *Chapter 6* we present briefly some extensions and competitor models such as Nonlinear $k - \epsilon$, Algebraic R_{ij}, $R_{ij} - \epsilon$ and Renormalization Group RNG-$k - \epsilon$ models.

The compressible $k - \epsilon$ model is treated in Part II, (*Chapters 7,8,9)* in a similar fashion as Chapter 3 to 6, except that Favre's weighted average is used. Of course the number of hypothesis to derive the model is much greater, so much so that all the details for the derivation of the ϵ equation in the model are not given. However, algorithms and stability are treated in depth. *Chapter 10* is devoted to numerical results.

Finally in Part III, convection by a turbulent velocity field is studied from a mathematical point of view(*Chapter 11*). Homogenization and multiple scales expansions are introduced. Its generalization to the incompressible Navier-Stokes equations is presented in *Chapter 12* . It justifies some of the assumptions of the $k - \epsilon$ model and it shows the limits of the model.

In summary this book is not about fluid mechanics but about numerical analysis; it does not attempt to justify the turbulence models on physical grounds but on mathematical ones. Thus the reader will not find many assessments about the qualities of the model for predicting turbulent flow on particular configurations. However, the book states that if certain hypotheses are met, if proper boundary conditions are applied and if the numerical algorithms are correctly chosen then the $k-\epsilon$ model is capable of giving reasonable answers because its derivation is not entirely heuristic.

CHAPTER 1

THE NAVIER-STOKES EQUATIONS

1. INTRODUCTION

This chapter contains a brief presentation of the equations for compressible and incompressible flows and some comments on the boundary layer equations and on wall laws. For more details the reader is sent to Anderson [1984], Bachelor [1970], Cousteix[1990], Landau-Lifchitz [1953], Panton [1984], Peyret[1985], Pironneau[1989]. Turbulence is discussed from the point of view of existence, uniqueness and dynamical systems results for the Navier-Stokes equations.

2. GENERAL EQUATIONS OF FLUIDS DYNAMICS

2.1 Conservation of mass

To study the motion of a fluid which occupies a domain Ω of R^3 over a time interval [0,T'], we shall denote by O any regular subdomain of Ω , by x any point of Ω and by t an instant of time. The fluid is characterized by

- the density field $\rho(x,t)$,
- the velocity vector field $u(x,t)$,
- the pressure field $p(x,t)$,
- the temperature field $T(x,t)$.

To conserve mass, the rate of change of mass of fluid in O, $\partial_t \int_O \rho$, has to be equal to the mass flux, $-\int_{\partial O} \rho u.n$, across the boundary ∂O of O, (n denotes the exterior normal to ∂O).

By using Stokes' formula

$$\int_O \nabla.(\rho u) = \int_{\partial O} \rho u.n$$

and the fact that O is arbitrary, the *equation of conservation of mass* is found:

$$\partial_t \rho + \nabla.(\rho u) = 0.$$

It is also called the *Continuity equation.*

2.2 Conservation of momentum

Newton's law must be written for a volume element O of fluid.
By definition of u, a particle of the fluid at position x at time t will be approximately at $x + u(x,t)\delta t$ at time $t + \delta t$; its acceleration is therefore

$$\lim_{\delta t \to 0} \frac{1}{\delta t}[u(x + u(x,t)\delta t, t + \delta t) - u(x,t)] = \partial_t u + \sum_j u_j \partial_j u \equiv \partial_t u + u\nabla u.$$

The forces on O are:

• The external forces (electromagnetism, Coriolis, gravity...), equal to $\int_O$ f, if f is the force per unit volume.
• The pressure force and the viscous force due to the motion in the fluid, equal to $\int_{\partial O}(\sigma - p\mathbf{I})n$ where σ is the stress tensor , $\mathbf{I}$ is the unit tensor and n denotes the unit outer normal at ∂O .

Newton's laws are, therefore, written as: for all O,

$$\int_O \rho(\partial_t u + u\nabla u) = \int_O f - \int_{\partial O}(pn - \sigma n) = \int_O (f - \nabla p + \nabla.\sigma),$$

where the second equality comes from Stokes' formula.
Thus

$$\rho(\partial_t u + u\nabla u) + \nabla p - \nabla.\sigma = f.$$

Remark
By the continuity equation, this equation is equivalent to

$$\partial_t(\rho u) + \nabla.(\rho u \otimes u) + \nabla.(p\mathbf{I} - \sigma) = f.$$

Proof: develop the derivatives and use the equation of conservation of mass.

To proceed further an hypothesis is needed to relate the stress tensor σ with the velocity of the fluid.
The hypothesis of *Newtonian flow* is a linear law relating σ to ∇u :

$$\sigma = \mu(\nabla u + \nabla u^T) + (\xi - \frac{2\mu}{3})\mathbf{I}\nabla.u$$

μ and ξ are the first and second viscosities of the fluid. For air and water the second viscosity ζ is very small; in this book it is assumed that $\zeta = 0$.

With this definition for σ, *equation of conservation of momentum* is found:

$$\partial_t(\rho u) + \nabla.(\rho u \otimes u) + \nabla p - \nabla.[\mu(\nabla u + \nabla u^T) - \frac{2\mu}{3}\mathbf{I}\nabla.u] = f.$$

It is easy to verify that the equation can also be written as

$$\rho(\partial_t u + u\nabla u) - \mu\Delta u - \frac{\mu}{3}\nabla(\nabla.u) + \nabla p = f$$

because $\nabla.\nabla u = \Delta u$ and $\nabla.\nabla u^T = \nabla(\nabla.u)$. Taking into account the continuity equation, the previous equation can be rewritten in *conservative form* as

$$\partial_t(\rho u) + \nabla.(\rho u \otimes u) + \nabla p - \mu\Delta u - \frac{\mu}{3}\nabla(\nabla.u) = 0.$$

2.3 Conservation of energy and the state equation

Finally, an equation called *conservation of energy* can be obtained by writing the total energy of a volume element *O(t)* moving with the fluid. Recall that the energy E is the sum of the work done by the forces and the amount of heat received. The energy in a volume element O has two origin:

- The internal energy e.
- The kinetic energy $u^2/2$.

So the energy in O is

$$\int_O \rho E \text{ where } E = \frac{u^2}{2} + e.$$

The work done by the forces is the integral over O of $u.(f+\sigma-p\mathbf{I})n$. By definition of the temperature T, if there is no heat source (combustion...) then the amount of heat received (lost) is proportional to the flux of the temperature gradient, i.e. the integral on ∂O of $\kappa\nabla T.n$. The constant κ is called the *thermal diffusivity* . So the following equation is obtained:

$$\frac{d}{dt}\int_{O(t)}\rho E=\int_O\{\partial_t\rho E+\nabla.[u\rho E]\}=\int_O u.f+\int_{\partial O}[u(\sigma-p\mathbf{I})-\kappa\nabla T]n$$

With the continuity equation and Stokes' formula, *the energy equation* is obtained:

$$\partial_t[\rho E]+\nabla.(u[\rho E+p])=\nabla.(u\sigma+\kappa\nabla T)+f.u.$$

For an *ideal fluid* , such as air and water in non extreme situations, C_v and C_p being physical constants, we have

$$e=C_vT \text{ and } E=C_vT+\frac{u^2}{2},$$

and the *equation of state*

$$\frac{p}{\rho}=RT,$$

where R is an ideal gas constant. With $\gamma=C_p/C_v=R/C_v+1$, the above can be written as follows:

$$e=\frac{p}{\rho(\gamma-1)}.$$

With the definiton of σ, the equation for E becomes

$$\partial_t[\rho\frac{u^2}{2}+\frac{p}{\gamma-1}]+\nabla.\{u[\rho\frac{u^2}{2}+\frac{\gamma}{\gamma-1}p]\}$$
$$=\nabla.\{\kappa\nabla T+[\mu(\nabla u+\nabla u^T)-\frac{2}{3}\mu\mathbf{I}\nabla.u]u\}+f.u$$

The energy equation can also be written in terms of the temperature

$$\partial_t T + u\nabla T + (\gamma - 1)T\nabla.u - \frac{\kappa}{\rho C_v}\Delta T = \frac{1}{\rho C_v}[|\nabla u + \nabla u^T|^2 \frac{\mu}{2} - \frac{2\mu}{3}|\nabla.u|^2].$$

By introducing the *entropy* :

$$s \equiv \frac{R}{\gamma - 1} \log \frac{p}{\rho^\gamma},$$

another form of the energy equation is (Cf. Landau-Lifchitz [141 p. 236]):

$$\rho T(\partial_t s + u\nabla s) = \frac{\mu}{2}|\nabla u + \nabla u^T|^2 - \frac{2}{3}\mu|\nabla.u|^2 + \kappa\Delta T.$$

Remark

Some values for the physical constants.
For air: $\rho = 1.2 \times 10^{-3} g/cm^3$, $\mu = 1.8 \times 10^{-4} g/cm.s$, $\kappa = 0.2 cm^2/s$, $\gamma = 1.4$, $R = 2.87 \times 10^6 cm^2/s^2.^oC$,
and for water: $\rho = 1 g/cm^3$, $\mu = 0.01 g/cm.s$, $\kappa = 1.4 \times 10^{-3} cm^2/s$, $\gamma = 1$, $R = 0.2410 cm^2/s^2.^oC$.

2.4 Boundary conditions

There are all kind of boundary conditions possible for the Navier-Stokes equations, (and also very few proofs that they work). The most popular set is, u, E, ρ given at time zero and

- u given
- T given, or its normal derivative.
- ρ given on the points x of Γ where $u(x).n(x) < 0$.

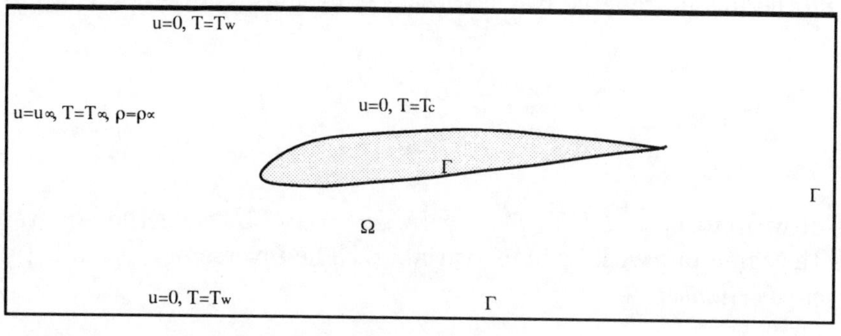

Figure 1.1 : A possible set of boundary conditions for a wing in a wind tunnel .

3. INCOMPRESSIBLE FLOWS

When ρ is practically constant (water for example or air at low velocity) we can neglect its derivatives. Then the general equations become the incompressible *Navier Stokes equations*

$$\nabla.u = 0,$$

$$\partial_t u + u\nabla u + \nabla p - \nu\Delta u = f,$$

where $\nu = \mu/\rho$ is the *kinematic viscosity* of the fluid and $p \to p/\rho$ and $f \to f/\rho$ are the reduced pressure and external force.
An equation for the temperature T can be obtained from the compressible one by assuming ρ constant

$$\partial_t T + u\nabla T - \frac{\kappa}{\rho C_v}\Delta T = \frac{\nu}{2C_v}|\nabla u + \nabla u^T|^2.$$

4. THE REYNOLDS NUMBER

Let us come back to the system for incompressible flows and let us rewrite it in non dimensional form.
Let U be a characteristic velocity of the flow under study (for example one of the non homogeneous boundary conditions).
Let L be the characteristic length (for example the diameter of Ω) and T_1 a characteristic time (which is a priori equal to L/U). Let us put

$$u' = \frac{u}{U}; \quad x' = \frac{x}{L}; \quad t' = \frac{t}{T_1}.$$

Then the incompressible Navier-Stokes equations can be rewritten as

$$\nabla_{x'}.u' = 0$$

$$\frac{L}{T_1 U}\partial_{t'}u' + u'\nabla_{x'}u' + U^{-2}\nabla_{x'}p - \frac{\nu}{LU}\Delta_{x'}u' = f\frac{L}{U^2}.$$

So, if we put $T_1 = L/U$, $p' = p/U^2$, $\nu' = \nu/LU$, then the equations are the same but with "prime" variables. The inverse of ν' is called the *Reynolds number.*

$$Re = \frac{UL}{\nu.}$$

Some examples: for a glider of characteristic length one meter crusing at speed one meter per second, $Re \approx 7 \times 10^4$. For cars running at 3m/s, $Re \approx 6 \times 10^5$ and for airplanes at 30m/s, $Re \approx 2 \times 10^7$.
When $\nu' >> 1$ (bio-fluid mechanics or very viscous flows), $\nu'\Delta u'$ dominates $u'\nabla u'$ and $u'_{t'}$; inertial terms can be neglected and the *Stokes problem* is obtained:

$$-\nu\Delta u + \nabla p = f, \quad \nabla . u = 0$$

5. MATHEMATICS AND TURBULENCE

5.1 Existence and uniqueness

The *compressible Navier-Stokes equations,* with suitable boundary and initial conditions, is complete in the sense that it has d+3 unknown scalars (ρ, u, p, T) and d+3 scalar equations.
Matsumura -Nishida[1980], Valli [1981], Lions[1993] have shown that the system is well posed with regular initial conditions (ρ^0, u^0, p^0, T^0) satisfying the law of perfect gas and regular boundary conditions : u, T given on the boundary, p calculated from the law of perfect gas and ρ given on the part of the boundary in which $u.n < 0$.

The incompressible Navier-Stokes equations are well posed in $Q = \Omega \times]0, T'[$ where Ω is a regular open bounded set in R^d, d = 2 or 3, with the following initial and boundary conditions :

$$u(x, 0) = u^0(x) \quad x \in \Omega$$

$$u(x, t) = u_\Gamma(x) \quad x \in \Gamma, \quad t \in]0, T'[.$$

The problem is reframed in the following spaces

$$J(\Omega) = \{v \in H^1(\Omega)^n : \nabla . v = 0\}$$

$$J_o(\Omega) = \{u \in J(\Omega) : u|_\Gamma = 0\}.$$

The problem is to find u $\in L^2(0, T', J(\Omega)) \cap L^\infty(0, T'; L^2(\Omega))$ such that

$$(\partial_t u, v) + \nu(\nabla u, \nabla v) + (u\nabla u, v) = (f, v), \quad \forall v \in J_o(\Omega)$$

$$u(0) = u^0, \quad u - u_\Gamma \in L^2(O, T', J_o(\Omega))$$

In the above $(\cdot, \cdot)$ denotes the scalar product in $L^2(\Omega)$; the equality is in the L^2 sense in t. The data must be smooth in the sense $f \in L^2(Q)$ and

u^0, $\tilde{u_\Gamma}$ in $J(\Omega)$, where the tilde denotes an extention in Ω of a function defined on Γ. The hypothesis, u in $L^\infty(O,T';L^2(\Omega))$ is to insure that the integrals containing $u\nabla u$ exist.

Theorem 1
The Incompressible Navier-Stokes equations in variational form have at least one solution. In two dimensions, the solution is unique; if f=0 and $|u^0|_1$ *is small or if u is smooth* (i.e. $u \in L^\infty(O,T';L^4(\Omega))$ *then the solution is unique in three dimensions.*

Proof: See Lions [1968].

5.2. Regularity of the solution

It should be emphasized that the last statement of the theorem says that if u is smooth, it is unique; but one cannot in general prove this type of regularity for the solution constructed in theorem 1; so its uniqueness is an open problem. There are important reports which deal with the study of possible singularities in the solutions. For instance, it is known that the singularities are "local " (Cafarelli et al. [1982], Constantin et al. [1985]) and and that their Hausdorff dimension is less than 1.

The Hausdorff dimension [1] d of a set O is defined as a limit (if it exists) of log $N(\epsilon)/log(1/\epsilon)$ when $\epsilon \longrightarrow 0$, where $N(\epsilon)$ is the minimum number of cubes (or other) having the length of their sides less or equal to ϵ and which cover O. ($d = 0$ if O is a point, $d = 1$ if O is a curve of finite length, $d = 2$ for a surface... non integer numbers can be found in fractals (Feder [1988])).

So uniqueness for the Navier-Stokes equations is an open problem and it is related to regularity of solutions and to continuous dependence on the initial data. Since this book deals with "turbulence" it is interesting to remark that turbulence is essentially a three dimensional phenomenon and that it is at least a loss of continuous dependence upon initial data for the mean flow.

5.3. Long time behaviour

In general we are interested in the solution for large times. In practice the flow does not seem highly dependent upon initial conditions: the flow around a car for instance does not really depend on its acceleration history.

1 This is Kolmogorov's definition.

There are many ways "to forget" these initial conditions for a flow; here are two examples:

1 . the flow converges to a *steady state* independent of t;
2 . the flow becomes *periodic* in time.

Moreover, the possibility is not excluded that a little "memory" of initial conditions still remains; thus in 1 the stationary state could depend upon initial conditions since the stationary Navier-Stokes equations have many solutions when the Reynolds number is large. We distinguish other limiting states:

3 . *Quasi-periodic* flows: the Fourier transform $t \to |u(x,t)|$ (where x is a fixed but arbitrary point of the domain Ω) has a discrete spectrum and two frequencies at least are not commensurable.
4 . *Chaotic* flows with strange attractors: $t \to |u(x,t)|$ have a continuous spectrum and the Poincaré sections (for example the points $\{u_1(x,nk), u_2(x,nk)\}_n$ for a given x and k) have dense regions of points filling a complete zone of space (figure 1.2). In case 1, the Poincaré sections are reduced to a point when n is large, in cases of 2 and 3, the points are on a curve.

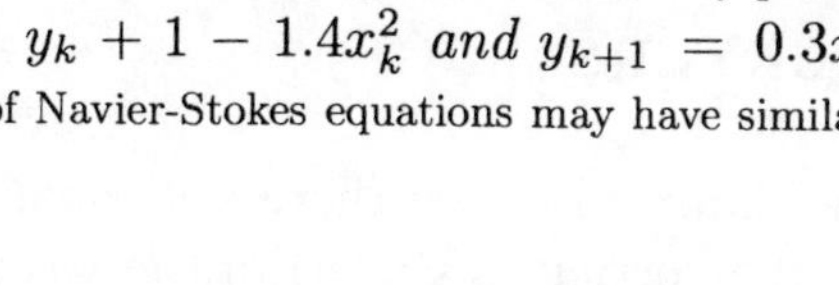

Figure 1.2 : Henon's attractor. This figure is obtained by plotting $\{x_k, y_k\}_{k=1,2,...}$ obtained by $x_{k+1} = y_k + 1 - 1.4x_k^2$ *and* $y_{k+1} = 0.3x_k$. Poincaré maps of "turbulent" solutions of Navier-Stokes equations may have similar features (Bergé et al [1984]).

In fact, experience shows that most flows pass through the 4 regimes, in the order 1 to 4, when the Reynolds number Re $\equiv 1/\nu$ (ν is the effective viscosity) increases and the change of regime takes place at the bifurcation points of the mapping $\nu \to u$, where u is a stationary solution of the Navier-Stokes equations. As in practice Re is very large (for us ν is small), regime 4 dominates; thus we may take this as a *mathematical definition of Turbulence* . The following points are under study:

• Whether there exists attractors, and if so, can we characterize any of their properties? (Hausdorff dimension, inertial manifold....cf. Ruelle-Takens[1971], Ghidaglia [1986], Foias et al.[1985] or Bergé et al [1984] and the bibliography therein).
• Does $u(x,t)$ behave in a stochastic way and if so, by which law? Can we deduce some equations for average quantities such as $\overline{u}$, $\overline{|u|^2}$, $\overline{|\nabla \times u|^2}$... this is the problem of turbulence modeling.

Here are the main results related to attractors for Navier-Stokes equations.
Consider the incompressible Navier-Stokes equations with $u_\Gamma = 0$, f independent of t, and Ω a subset of R^2. This system has an attractor whose Hausdorff dimension is between $cRe^{4/3}$ and CRe^2 where $Re = \sqrt{f}\mathrm{diam}(\Omega)/\nu$ (cf. Constantin and al.[1985], Ruelle [1980][1984]). These results are interesting because they give an upper bound for the number of points needed to calculate such flows (this number is therefore at least proportional to $\nu^{-9/4}$).
In three dimensions, it is not known that the incompressible Navier-Stokes equations with the same boundary conditions has an attractor but if an attractor exists and is (roughly) bounded by M in $W^{1,\infty}$, then its dimension is less than $CM^{3/4}\nu^{-9/4}$ (cf.Constantin et al[1985]).

Even if these results are refined the problem of turbulence modeling remains because these upper bounds for the number of computational points (or degrees of freedom) are too large for practical applications, as we shall see in the next chapter.

6. BOUNDARY LAYERS

When the Reynolds number is large there are usually strong gradients for the velocity and the vorticity ($\nabla \times u$) in the viscinity of solid walls. These are also rather difficult to simulate numerically. We must recall here some generalities on boundary layers. For clarity we limit the discussion to incompressible flows and send the reader to Cousteix [1990] for the general case.
The purpose if this section is to
• give the classical simplifications done to the Navier-Stokes equations in the boundary layers
• show that their thickness are of the order of $\sqrt{\nu}$.

6.1 Prandtl's laminar boundary layer equations

If ν is set to zero in the incompressible Navier-Stokes equations, what remains is the so called Euler equations:

$$\partial_t u + u\nabla u + \nabla p = 0, \quad \nabla.u = 0.$$

Standard boundary conditions for these are

$$u.n|_\Gamma = u_\Gamma.n, \quad \nabla \times u|_{\Gamma^-} = \omega_\Gamma$$

where $\Gamma^- = \{x \in \Gamma : \quad u(x).n(x) < 0\}$.
This difference in boundary conditions creates a thin layer in which the flow passes from the Euler regime into the Navier-Stokes regime. Also since

$$u\nabla u = -u \times \nabla \times u + \frac{1}{2}\nabla(u.u) \text{ and } -\Delta u = \nabla \times \nabla \times u - \nabla(\nabla.u),$$

there is a simple stationary solution (*potential flow*) to the Euler equation with $\nabla \times u = 0$

$$u = \nabla\phi, \quad \Delta\phi = 0, \quad p = -\frac{1}{2}u^2.$$

Consider the bidimensional stationary case; let x be the coordinate parallel to Γ and y the normal one; similarly u_1, u_2 denotes the x and y component of u; it can be shown that near a solid wall some terms can be dropped in the Navier-Stokes equation. The following describes correctly the flow near the boundaries as long as u_1 does not change sign (Prandtl's equations without rescaling):

$$\begin{aligned} &u_1\partial_x u_1 + u_2\partial_y u_1 - \nu\partial_{yy}u_1 = -\partial_x p, \\ &\partial_y p = 0, \\ &\partial_y u_2 = -\partial_x u_1. \end{aligned}$$

These would be valid in a thin layer B_δ above the wall Γ. In $\Omega_\delta = \Omega - B_\delta$, Euler equations or its potential approximation are valid. Matching the velocities from the two regions at the interface Σ is necessary and provides the reduced equations in B_δ with a boundary condition above the walls.

6.2 The Falkner-Skan equations

In 2 dimensions $\nabla.u = 0$ implies that there exists a *stream function* ψ such that $u = \{\partial_y\psi, -\partial_x\psi\}$ and so Prandtl's equations become,

$$\frac{1}{2}\partial_x(\partial_y\psi)^2 - \partial_x\psi\partial_{yy}\psi - \nu\partial_{yyy}\psi = U\partial_x U$$

where U is the solution of Euler equations. Boundary conditions are $\psi(x,0) = \partial_y\psi(x,0) = 0$, $\partial_y\psi(x,\delta) = U$.

6.3 Flat plate

Assume that, for some m:

$$\psi(x,y) = x^m f(yx^{m-1}).$$

By replacing ψ in the Falkner-Skan equation, the following is found:

$$(2m-1)(f')^2 - mff'' - \nu f''' = U\partial_x U x^{3-4m}$$

Thus there is a self similar solution when $U|_\Gamma = Cx^\beta$ with $m = (1-\beta)/2$. For the flat plate, $\beta = 0$, so $m = 1/2$. In such case, the computations reduce to solving an O.D.E; it shows that the behaviour of u near y=0 is like $U(1-e^{-y/\sqrt{\nu}})$.

7. WALL LAWS

The previous equations indicate that in the viscinity of walls, the velocity passes from 0 to O(1) over a distance $\delta = O(\sqrt{\nu})$. Numerical simulations will have to take this fact into account by refining the mesh accordingly in the boundary layers. Wall laws are an attempt to remove this constraint. Although they have scored so far limited success they are important in turbulence modeling.

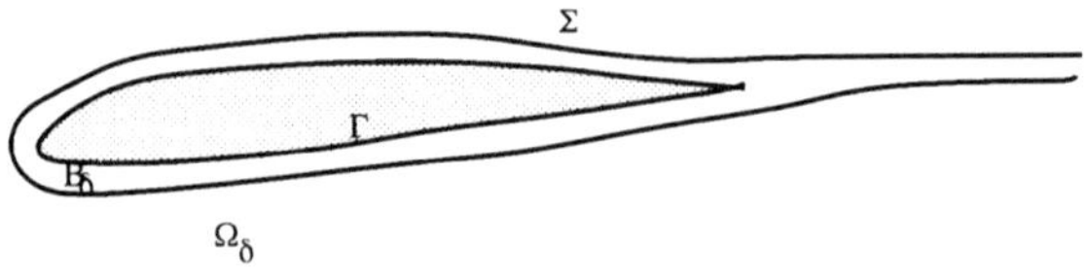

Figure 1.3 : An artificial boundary Σ surrounds the airfoil Γ (not counting the wake). The domain between Σ and Γ is B_δ and the domain exterior to Σ *is* Ω_δ.

7.1 The basic idea

The basic idea is to remove boundary layers from the computational domain (figure 1.3). Let $\delta(x)$ be the boundary layer thickness above Γ and let

$$B_\delta = \{x - n(x)\lambda : \quad x \in \Gamma, \quad \lambda \in]0, \delta(x)[\}.$$

The computational domain is now $\Omega_\delta = \Omega - B_\delta$ and the new boundary $\Sigma = \partial\Omega_\delta$ replaces Γ.
Of course we need a boundary condition for u on Σ. One possibility is to use a Taylor expansion of u at $x' + \delta(x')n(x')$ which is a point of Γ when $x' \in \Sigma$:

$$u(x' + \delta(x')n(x')) = u(x') + \delta(x')\frac{\partial u}{\partial n}(x') + o(\delta).$$

Therefore $u|_\Gamma = 0$ implies

$$u + \delta\frac{\partial u}{\partial n} \approx 0 \text{ on } \Sigma.$$

7.2 Wall laws for a rough boundary

However this works only if $\partial_{nn}u$ is smooth in B_δ, for instance in the part of the boundary layer far from the leading edge of a wing.
Boundary layers do not satify this property everywhere; so another technique must be used namely a multiple scale expansion (homogenization). The method is easy to understand when it is used to account for wall roughtness.
So consider an aerodynamic wing profile Γ with very fine periodic irregularities on the surface (figure 1.4).

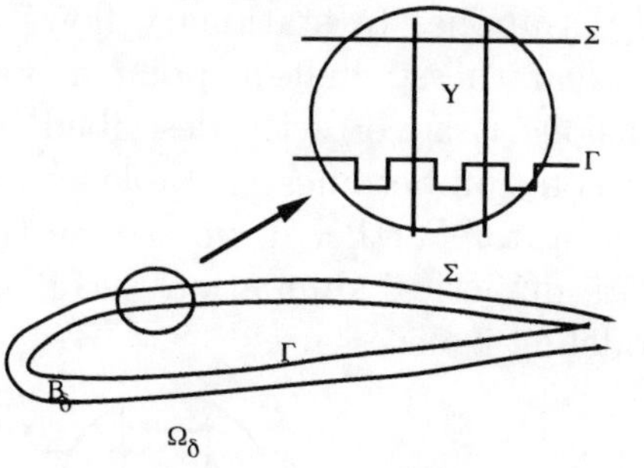

Figure 1.4 : An airfoil with rough surface. The boundary condition is applied on Σ rather than on Γ. To find the new boundary condition, an auxiliary problem is solved on a representative cell Y with periodic boundary conditions on the vertical sides.

Consider a flat surface Σ just above Γ. One seeks a boundary condition on Σ which has the same effect. Following Carrau[1993] and Le Tallec we assume that the flow is stationary and we consider two regions again, B_δ below Σ, i.e. between Γ and Σ, and Ω_δ above Σ. Because B_δ is thin it is conceivable that the flow is somewhat x_1- periodic below Σ. Let Y be a cell of periodicity. If $U_\Sigma \equiv u_1|_\Sigma$ was known the flow below Σ would be solution of Navier-Stokes equations,

$$u\nabla u + \nabla p - \nu \Delta u = 0 \quad \nabla . u = 0 \text{ in } Y,$$

$$u|_\Gamma = 0 \quad u_1|_\Sigma = U_{1\Sigma} \quad u_2|_\Sigma = 0, \quad u, p \;\; x_1 - \text{ periodic.}$$

The condition $u_2|_\Sigma = 0$ means that Σ is a stream line, which is true as a first approximation.
The solution of this problem depends nonlinearly on U_Σ. So $g(U_\Sigma) \equiv \partial u_1/\partial n|_\Sigma$ is some nonlinear function of U_Σ. The function $u \to g(u)$ can be tabulated by solving the problem in Y several times with different values of U_Σ.
Above Σ, u is also solution of Navier-Stokes equations so matching u and ∇u on Σ requires the following boundary conditions:

$$\frac{\partial u_1}{\partial n} = g(u_1), \quad u_2 = 0 \text{ on } \Sigma.$$

This is called a wall law. In effect, it removes the regions of strong gradients from the computational domain at the expense of a more complicated boundary condition.
This method can be justified by homogenization in the case of Stokes equations (Conca[1989]).

7.3 Wall laws for Turbulent boundary layers

The previous analysis applies to stationary flows. Consider now the case of an airfoil. Near the stagnation point a boundary layer develops. In two dimensions it is correctly described by the Falkner-Skan equation. Then the solution becomes unstable and there is a transition to a time dependent state. Further down stream the solution becomes fully "turbulent", its numerical simulation becomes almost impossible and turbulence modeling is necessary.

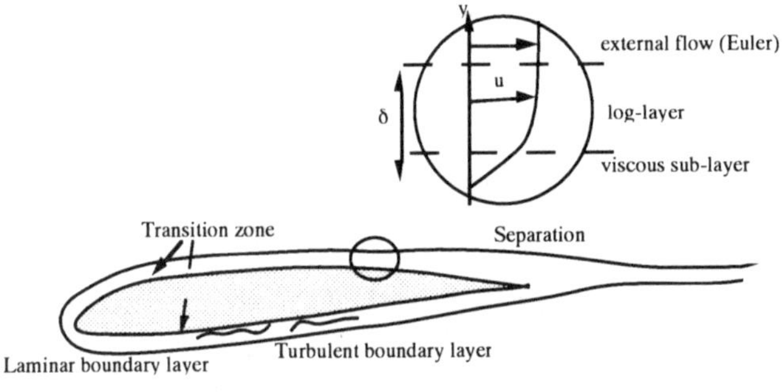

Figure 1.5 : Boundary layers have usually three zones: a laminar zone, followed by a transition zone and a turbulent zone. Boundary layers may separate if the airfoil is not sufficiently aerodynamic. In the turbulent part the horizontal velocity has a universal profile which is linear in the so called viscous sublayer and logarithmic in the log-layer.

Experiments show some universality in the behavior of turbulent attached boundary layers.
More precisely if x_1 denotes the direction of the mean flow parallel to the wall, and x_2 is the coordinate in the normal direction, define

$$u^* = \sqrt{\nu \frac{\partial U_1}{\partial x_2}}|_{x_2=0}, \quad y^* = \frac{\nu}{u^*}$$

where U is the time averaged velocity:

$$U(x) = \lim_{T\to\infty} \frac{1}{T}\int_0^T u(x,t).$$

Finally let

$$y^+ = \frac{y}{y^*}, \text{ and } u^+ = \frac{U}{u^*}.$$

Then experiments show that when $20 \le y^+ \le 100$, in the so called *logarithmic layer*, there is a universal formula for the scaled mean flow u^+,

$$u^+ = \frac{1}{\chi}\log y^+ + 5.5 \qquad \chi = 0.41.$$

The constant χ is the von Karman constant.
Nearer to the wall, $0 \le y^+ \le 20$, the mean flow is almost linear in y^+. This is the *viscous sublayer*

$$u^+ = y^+.$$

These formulae are nonlinear relations between u and $\partial u/\partial n$; they can be used to establish wall laws in place of the auxiliary Navier-Stokes equations of §7.2 with periodic conditions in Y. Usually one seeks a δ so that Σ is in the logarithmic layer, and then solves the Navier-Stokes equations with the following boundary conditions, called a wall-law:

$$u.n = 0,$$

$$\frac{u.s}{\sqrt{\nu|\frac{\partial u}{\partial n}|}} - \frac{1}{\chi}\log(\delta\sqrt{\frac{1}{\nu}|\frac{\partial u}{\partial n}|}) + \beta = 0,$$

where n is the normal and s a tangent to Σ, $\chi = 0.41$, $\beta = 5.5$.
Parès[1988] showed that the Navier-Stokes equations are well posed with these boundary conditions.

Remark
It is necessary to verify, a posteriori, that

$$20\sqrt{\nu|\frac{\partial u}{\partial n}|^{-1}} \leq \delta \leq 100\sqrt{\nu|\frac{\partial u}{\partial n}|^{-1}}.$$

Remark
Unfortunately boundary layers are also known to *separate*. This means that the sign of $\partial u_1/\partial n$ changes, Prandtl's analysis is no longer valid and the region of strong gradients is no longer thin. Then wall laws fail.

8. CONCLUSION

The Navier-Stokes equations which describe the motion of a Newtonian fluid are well posed up to a critical time t^*. Even in the incompressible case there is no global uniqueness result valid for all times independently of the Reynolds number. Some relate this lack of results to the existence of turbulence in fluids, because mathematically turbulence could be defined as a loss of regularity leading to a loss of continuous dependence of the solution upon its initial data.

For numerical simulations turbulence is not the only difficulty, strong gradients are also a serious difficulty. These are present in turbulent flows but also in boundary layers, even laminar boundary layers. However boundary layers are much better understood than turbulence and their modeling by wall laws is easier.

CHAPTER 2

HOMOGENEOUS INCOMPRESSIBLE TURBULENCE

1. INTRODUCTION

For incompressible flows, the Navier-Stokes equations, with a set of initial and boundary conditions are :

$$\partial_t u + u\nabla u + \nabla p - \nu\Delta u = f,$$
$$\nabla . u = 0,$$
$$u|_{t=0} = u^0,$$
$$u|_{\partial\Omega} = u_\Gamma.$$

These hold over the domain Ω occupied by the fluid, during a given interval of time $]0, T'[$. The data are :
• the external forces f,
• the reduced viscosity $\nu = \mu/\rho$ or the effective viscosity $\nu = Re^{-1}$ if u=0(1), $|\Omega| = 0(1)$.
• the initial conditions u^0,
• the boundary conditions u_Γ.
A well known experiment consists in studying the flow in a pipe behind a grid placed perpendicular to the flow across the pipe. When the grid is fine compared to the diameter of the pipe, velocity measurements at a point x_0 show that :
• these are noisy functions of time but have well defined mean values and moments,
• do not depend much upon the initial state u^0,
• do not depend much upon x_0 within a certain range.
To understand this chapter one must remember that turbulence was analyzed first from experiments. Experimental results are often in the form of one dimensional plots of physical quantities, like the temperature

at one point as a function of time, or two or three dimensional pictures of instantanuous velocity or pressure fields. When these "look" like random processes, then the flow is said to be turbulent. It is not easy to give a precise definition of these notions. Nevertheless we shall try via Fourier transforms. But before, let us recall that certain quantities, called invariants, cannot be fluctuating.

2. INVARIANTS

Consider the case $\Omega = R^3$ (no boundaries, flow quantities decay to zero at infinity) driven by initial conditions (no external forces).

2.1 The mean flow is an invariant

Indeed, integrated over O, a ball of radius r, the momentum equation for u gives

$$\int_O \partial_t u + \int_O u\nabla u + \int_O \nabla p - \int_O \nu\Delta u = 0$$

but

$$\int_O u\nabla u = \int_O \nabla.(u \otimes u) = \int_{\partial O} u(u.n);$$

$$\int_O \nabla p = \int_{\partial O} p.n; \text{ and } \int_O \nu\Delta u = \nu \int_{\partial O} \frac{\partial u}{\partial n}.$$

Since the boundary integrals tend to zero when $r \to \infty$ we have:

$$\partial_t \int_\Omega u = 0.$$

2.2 The mean kinetic energy is an invariant of Euler equations

Euler equations are obtained by setting $\nu = 0$. Multiplying the momentum equations by u and integrating in R^3 yields

$$\int_\Omega \partial_t u.u + \int_\Omega u\nabla u.u + \int_\Omega u\nabla p - \int_\Omega \nu\Delta u.u = 0,$$

$$\frac{1}{2}\int_\Omega \partial_t u^2 + \frac{1}{2}\int_\Omega u\nabla u^2 + \int_\Omega \nabla.(up) + \int_\Omega \nu|\nabla u|^2 = 0,$$

$$\frac{1}{2}\frac{\partial}{\partial t}\int_\Omega u^2 + \int_\Omega \nu|\nabla u|^2 = 0.$$

Note that kinetic energy becomes viscous dissipation. Notice also that if u^0 is smooth then $\nu|\nabla u^0|$ is small when ν is small. Thus in this case and for some time, the kinetic energy will be constant until the flow builds strong gradients.

2.3 Helicity is an invariant of Euler equations

Taking the curl of the momentum equation and denoting $\omega = \nabla \times u$, we find after multiplication by u

$$\int_\Omega \partial_t \omega u - \int_\Omega \nabla \times (u \times \omega) u - \int_\Omega \nu \Delta \omega u = 0;$$

i.e.

$$\frac{1}{2}\partial_t \int_\Omega \omega.u + \int_\Omega \nu\omega.\nabla \times \omega = 0,$$

because $u\nabla u = \frac{1}{2}\nabla u^2 - \nabla \times (u \times \omega)$, $\Delta\omega = -\nabla \times (\nabla \times \omega)$, and

$$\int_\Omega \partial_t \omega u = \int_\Omega (\partial_t \nabla \times u) u = \int_\Omega \partial_t u \nabla \times u = \int_\Omega \omega.\partial_t u.$$

Therefore when $\nu = 0$, helicity $\equiv \int_\Omega \omega.u$, is constant, so just like above for smooth initial conditions helicity will be quasi constant until the flow builds strong gradients.

2.4. Enstrophy is an invariant of Euler equations in 2D

Taking the curl of the momentum equation and multiplying by ω yields:

$$\int_\Omega \omega.\partial_t \omega - \int_\Omega u\nabla\omega\omega - \int_\Omega \nu\Delta\omega\omega = 0,$$

$$\frac{1}{2}\int_\Omega \partial_t \omega^2 - \int_\Omega \nu|\nabla \times \omega|^2 = 0.$$

When $\nu = 0$, enstrophy $\equiv \int_\Omega \omega^2$, is constant.

We notice that 4 quantities seem to play an important role in the balance of energy:

- kinetic energy $1/2 \int_\Omega u^2$,
- the rate of viscous dissipation $\int_\Omega \nu|\nabla u|^2$,
- helicity $\int_\Omega \omega.u$,
- enstrophy $\int_\Omega \omega^2$ in two dimensions.

2.5 Ergodicity

The ergodic theorem of probability theory (Neveu[1967]) says that, under certain conditions, statistical means can be replace by time averages or spatial averages in the case of multidimensional processes.

Turbulence is usually assumed ergodic, otherwise experiments are impossible to process. If it is so then the invariants defined above and scaled per unit volume can be defined also as,

- Turbulent kinetic energy: $\frac{1}{2}\langle|u|^2\rangle$,
- Rate of viscous dissipation: $\nu\langle|\nabla u|^2\rangle$,
- Helicity: $\langle\omega.u\rangle$,

3. FOURIER MODES

Consider the case $\Omega = R^3$. If u is square integrable the k-Fourier mode of u at time t is a vector of R^3 :

$$Fu \equiv U(k,t) = (\frac{1}{2\pi})^3 \int_\Omega u(x,t)e^{-ik.x}dx.$$

It is well defined for all $k \in R^3$. Conversely

$$F^{-1}U \equiv u(x,t) = \int_{R^3} U(k,t)e^{ik.x}dk.$$

The Fourier transform of Navier-Stokes equations is an integro differential system:

$$\partial_t U(k) + i\int_{p+q=k} U(p).qU(q) + ikP(k) + \nu|k|^2U(k) = F(k),$$

$$k.U(k) = 0,$$

where P, F are the Fourier transforms of p and f. Upon multiplication by $-ik$ and by making use of $k.U(k) = 0$, the first equation gives a formula for $P(k)$:

$$|k|^2P(k) = -ik.F(k) - k\int_{p+q=k} U(p).qU(q).$$

So one is left with ($(k\otimes k)_{ij} = k_ik_j$):

$$\partial_t U(k)+i(I-\frac{k\otimes k}{|k|^2})\int_{p+q=k} U(p).qU(q)+\nu|k|^2U(k) = (1-\frac{k\otimes k}{|k|^2})F(k).$$

We observe that

- D=(I- $k\otimes k/|k|^2$) is a projection operator on the space of divergence free functions in the sense that $F^{-1}(DFf)$ is divergence free.

- $\int_{p+q=k} U(p).qU(q)$ is an interaction operator between different Fourier modes. There are two types of interactions of interest to the physicist:
- $|p|, |q| >> |k|, p+q=k$ i.e. two small eddies can make a big one,
- $|p|, |q| = 0(|k|),\ p+q=k$ two eddies produce a third one of the same size.
- $\partial_t U(k) + \nu|k|^2 U(k) = 0 \Leftrightarrow U(k,t) = U(k,0)e^{-\nu|k|^2 t}$ so high modes $(k >> 1)$ have: $\nu|k|^2 >> 1$ and are damped faster.

4. STATISTICAL TURBULENCE

Periodic and quasi-periodic functions have a discrete or countable spectrum: $U(k) = 0$ except for a finite or countable number of vectors k.

Definition
If the spectrum of u (i.e. $\{U(k) \neq 0\}_k$) is dense in some large interval $]k_1, k_2[$, $k_2 >> k_1$ then we may say that the flow is turbulent .

The theory of dynamical systems identifies this property with the appearance of *chaos* , the loss of continuous dependence upon initial data and the random aspect of the solution.
The solution of Navier-Stokes equations is deterministic but it looks like a random process because from one experiment to the next it is not possible to reproduce $u(x,t)$ but only its mean over a time interval or a spatial average. Thus following Kraichnan[1972] let us consider the problem with random initial data $u^0(x,\omega)$. The solution is random and each realization may be denoted by $u(x,t,\omega)$. It has a probability density $d\omega$. Means, $\int u d\omega$, and moments, $\int \prod_{i=1,2,3} u_i^{p_i} d\omega$, of u can be evaluated, in principle ($u = \{u_1, u_2, u_3\}$).

4.1 Homogeneous Isotropic Turbulence

A turbulence is *homogeneous* if all the means and moments of u and their gradients are independent of x. A turbulence is *stationary* if the same are independent of t.
A turbulence is *isotropic* if all means and moments of u are equal to the same computed with $u(Mx,t)$ where M is a rotation matrix (invariance by rotation).

Moments. We may extend these definitions to *correlation functions* such as

$$R_{ij}(x,t,r) = \int u_i(x+r,t,\omega)u_j(x,t,\omega)d\omega,$$

and require from homogeneous isotropic turbulence to have R_{ij} and the like independent of x, t and function of $|r|$ only.

Properties of R

Isotropic homogeneous turbulent flows have all kinds of identities such as $R_{ij}(x,t,r) = R_{ij}(r) = R_{ij}(-r)$ (cf Batchelor[1970]). Furthermore, a direct consequence of the definitions is

$$\langle U_m(p)U_n(q)\rangle$$
$$= (\frac{1}{2\pi})^6 \int \langle u_m(x)u_n(y)\rangle e^{-ip.x-iq.y}dxdy = \Phi_{mn}(p)\delta(p+q)$$

where $\langle\cdot\rangle$ denotes the statistical mean and where Φ is the Fourier transform of R:

$$\Phi_{mn}(p) = (\frac{1}{2\pi})^3 \int R_{mn}(r)e^{-ip.r}dr.$$

Proposition

In homogeneous isotropic turbulence the tensor Φ has the following form

$$\Phi_{pq}(k) = \frac{E(|k|)}{4\pi|k|^2}(\delta_{pq} - \frac{k_p k_q}{|k|^2})$$

where $E(K)$ is a scalar function interpreted as the "kinetic energy density" of the turbulence.

4.2 Galilean invariance

Let U be the mean flow. The change $x \to x + Ut$, $u(x,t) \to u(x+Ut) - U$ does not change the Navier-Stokes equations, this is the so called Gallilean invariance. But now the new velocity has mean zero. Thus without loss of generality for homogeneous turbulence we may assume $\int u d\omega = 0$.

Of the quantities related to the invariants are left only :

- the kinetic energy density E,
- the rate of viscous dissipation $\epsilon = \langle\nu|\nabla u|^2\rangle$.

Remark

Note that

$$\Delta u = \nabla(\nabla.u) - \nabla\times\nabla\times u \quad \text{so} \quad \int_{R^3}|\nabla u|^2 = \int_{R^3}|\nabla\times u|^2.$$

If ergodicity allows to replace a statistical average by an ensemble average (cf chapter 3) then an alternative definition for ϵ is $\epsilon = \nu\langle|\nabla \times u|^2\rangle$.

5. THE KOLMOGOROV LAW

Consider an homogeneous isotropic stationary turbulence. Then the kinetic energy density E is a polynomial function of the modulus $K = |k|$ of the wave number k.

5.1 Conjecture (the $K^{-5/3}$ law)

When $U\nu^{-1} << K << \epsilon^{1/4}\nu^{-3/4}$, $K \to E(K)$ *decays like* $K^{-5/3}$. *More precisely*

$$E(K) \approx 1.5\epsilon^{\frac{2}{3}}K^{-\frac{5}{3}}.$$

Argument:
Assume that $E(K)$ is a function of ϵ and K. In terms of dimensions

$$\nu \approx UL, \quad \epsilon \approx \nu U^2L^{-2} \approx U^3L^{-1} \text{ and } R_{ii} \approx U^2, K \approx 1/L,$$

so $E(K) \approx LU^2$. Now when $K << U/\nu$, $E(K)$ cannot be function of ν because the viscosity is unable to damp out such low modes (see §2.4). The only quantities which we can use are K and ϵ. Therefore

$$E(K) \approx LU^2 \approx L(\epsilon L)^{\frac{2}{3}} \approx K^{-1}\epsilon^{\frac{2}{3}}K^{-\frac{2}{3}} \approx \epsilon^{\frac{2}{3}}K^{-\frac{5}{3}}.$$

Experiments show that the constant is around 1.5.

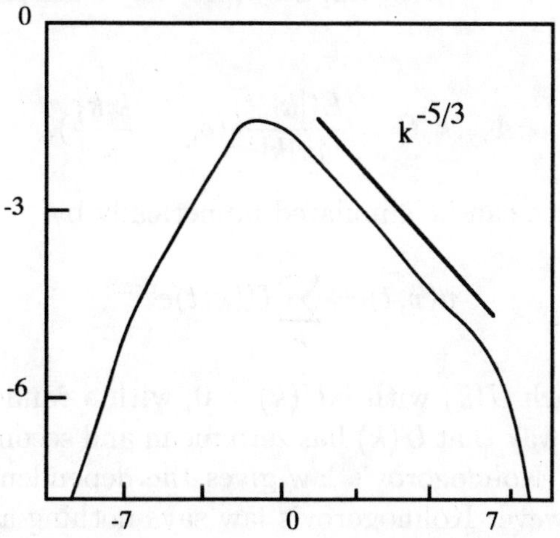

Figure 2.1 Typical spectrum, $\log_2 E(k)$ versus $\log_2 k$, for a turbulent flow in regions of homogeneous turbulence.

Initial conditions determine $E(K)$ at time t=0. Then there is a slow evolution so as a first approximation E can be considered as stationary. Viscosity affects the end of the spectrum. Initial conditions determine the beginning of the spectrum. In between there is a large range of K, called the *inertial range* , where the Kolmogorov law holds.

Viscosity acts when the eddy characteristic velocity U and its characteristic length L are such that $UL = 0(\nu)$. But $U \approx \epsilon^{1/3}K^{-1/3}$ so $UL \approx \epsilon^{1/3}K^{-4/3} = 0(\nu)$. Thus viscosity affects the part of the spectrum where $K = 0(\nu^{-3/4})$. This is called the *viscous sub-range.* This phenomenological argument leads to think that small eddies of size $0(\nu^{3/4})$ cannot live a long time because of the viscosity. Thus $E(K)$ has a compact [1] support, it does not extend to infinity.

Remark
With ϵ and K it is possible to make a time scale: $T \approx L/U \approx \epsilon^{-1/3}K^{1/3}K^{-1} = \epsilon^{-1/3}K^{-2/3}$

6. REPRESENTATION OF AN HOMOGENEOUS ISOTROPIC TURBULENCE

A gaussian random process is completely determined by its first and second order moments. Each mode $U(k)$ of a turbulent flow is not Gaussian but it is not so far from it either.
We have seen earlier that $\Phi_{ij}(k,t)$ is the Fourier transform of $R_{ij}(r,t)$. The constraint $\nabla.u = 0$ implies (see for example Lesieur[1987], Stanisic[1985])

$$\Phi_{ij}(k,t) = \frac{E(|k|,t)}{4\pi|k|^2}(\delta_{ij} - \frac{k_i k_j}{|k|^2}).$$

Thus a turbulence can be simulated numerically by

$$u(x,t) = \sum_k U(k,t)e^{ik.x},$$

and sampling each $U(k)$ with $k.U(k) = 0$, with a random number generator in such a way that $U(k)$ has zero mean and second order moment equal to $\Phi(k,t)$. Kolmogorov's law gives the dependence of E with respect to $|k|$. However Kolmogorov's law says nothing about the dependence on t.

---

[1] Not quite; intermitency is one exception...

Although isotropic turbulence is approximately stationary, experimental observation shows that in free turbulence ($f = 0$) E decays like

$$E \approx t^{-\frac{5}{4}}.$$

Kolmogorov's law also says nothing about the beginning of the spectrum ($K = O(1)$). For a more elaborated model for the dependence of E upon t the reader is sent to Kraichnan's EDQNM [2](Lesieur[1987] for example); the model allows to compute the evolution of $E(K, t)$ from an initial value $E^0(K)$, by solving an integro differential equation.

7. DIRECT SIMULATION

Small eddies exist up to $K = 0(\nu^{-3/4})$. So to capture them, we need a mesh size $h \approx \nu^{3/4}$. The total number of points in 3d is then $N = \nu^{-9/4}$. With $\nu^{-1} = 10^6$, a reasonable number for industrial applications, $N = 10^{13.5}$. At this time $N = O(10^6)$ is the technological maximum for supercomputers.
Computers may be able to handle $N = 10^9$ by the next century but this gives a Reynolds number around 10^4. Even if such simulations may be insufficient for practical applications, they are useful:

- to validate turbulence models,
- to simulate large eddies when they are not too much influenced by small ones.

7.1 An explicit Spectral Method

We have seen that the Fourier modes of the velocity of an incompressible fluid is solution of

$$\partial_t U(k) + i(I - \frac{k \otimes k}{|k|^2})F[(F^{-1}U).\nabla(F^{-1}U)] + \nu|k|^2 U(k) = 0$$

for all $k \in R^3$. When the physical domain is not R^3 but the cube $]-\pi,\pi[^3$, and we search for periodic solutions then we may restrict k to be made of integers and replace Fourier transforms by Fourier series. A discretization is obtained by taking a finite number of modes instead of all the integers for each components. Let

$$H = \{k \in R^3 : k_j \in [-N, N], k_j \text{ integer }\}.$$

[2] EDQNM=Eddy Damped Quasi Normal Markovian approximation

Let F_N be the truncated Fourier transform, i.e. $(F_N u)_k$ is the k-Fourier mode of u is $k \in H$ and

$$(F_N u)_k \equiv 0 \text{ if } k \notin H.$$

Because u is real, $U(k)$ and the complex conjugate part $\overline{U(-k)}$ are equal; this allows to reduce the number of unknown modes by half.

Similarly

$$F_N^{-1} U = \sum_{k \in H} U_k e^{ik.x}.$$

Following Orszag[1977], a numerical approximation is

$$\partial_t U_k + i(I - \frac{k \otimes k}{|k|^2}) F_N[(F_N^{-1} U).\nabla(F_N^{-1} U)] + \nu |k|^2 U_k = 0$$

for all $k \in H$.

Notice that the product $u \nabla u$ is performed in the physical space rather than the Fourier space for numerical efficiency.

A good numerical scheme for this equation is the Leap-frog finite difference scheme for the convection part and Crank-Nicolson's scheme for the diffusion part (see Ritchmyer-Morton[1967]):

$$\frac{1}{2\delta t}[U_k^{n+1} - U_k^{n-1}] + i(I - \frac{k \otimes k}{|k|^2}) F_N[(F_N^{-1} U^n).\nabla(F_N^{-1} U^n)] +$$

$$+ \frac{\nu}{2} |k|^2 (U_k^n + U_k^{n+1}) = 0 \quad k \in H$$

For more details the reader is sent to , Brachet[1992], Orszag[1977], Peyret et al[1985], Rose et al [1978]...

Some numerical results are given for the decay of an isotropic turbulence in the unit cube with periodic boundary conditions. vorticity in the flow is shown on figure 2.2.

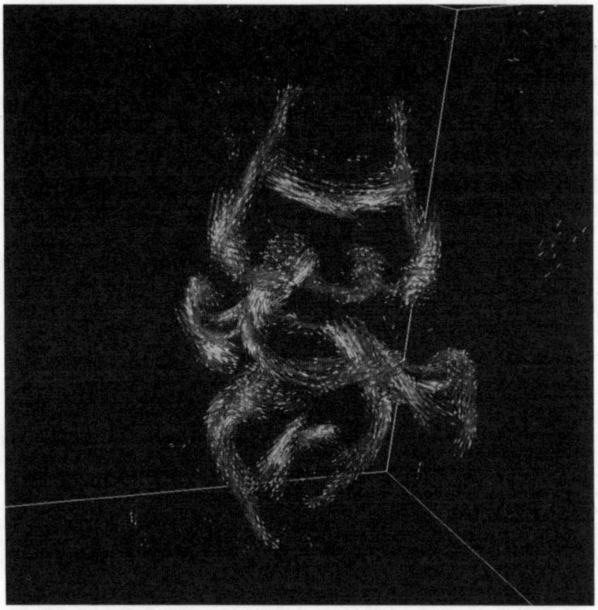

Figure 2.2 Vorticity tubes in a turbulence obtained by numerical integration of the Navier-Stokes equations Fourier transformed in space with 256^3 modes. (Courtesy of M. Brachet[1992])

8. SUBGRID-SCALE MODELING

Since much is known for the small eddies, via Kolmogorov's law, is it possible to use this knowledge in a computation?
One may split the mean part of the flow U from the oscillatory part u' and write

$$u(x,t) = U + u' \equiv \sum_{k<k_c} U_k(t)e^{ik.x} + \sum_{k\geq k_c} U_k(t)e^{ik.x}$$

The first part is computable on a grid of moderate size and the second part is computed by the random generator with $\epsilon(t)$ chosen so that the spectrum is continuous.
Such a program has been completed with success by Chollet et al[1980] [1984] with EDQNM for the time evolution of E between $n\delta t$ and $(n+1)\delta t$. Thus one can study the evolution of a turbulence which starts from u^0 at time zero and try to find approximate formula to represent the effect of the high modes (the second sum) on the low ones.

8.1 Turbulent diffusion

Experimentalists have since long noticed similarities between viscous flows and turbulent flows. Reynolds [1895] suggested that a turbulent flow is like a Newtonian flow with a non constant diffusion ν_T.
As we shall see in Chapter 3, when the decomposition $u = \bar{u} + u'$ is placed into the Navier-Stokes equations, it is found that

$$\partial_t \bar{u} + \bar{u}\nabla\bar{u} + \nabla P - \nu\Delta\bar{u} + \nabla.(\overline{u' \otimes u'}) = F, \quad \nabla.\bar{u} = 0.$$

Reynolds hypothesis assumes that

$$\nabla.(\overline{u' \otimes u'}) = -\nabla.(\nu_T \nabla\bar{u}).$$

For homogeneous turbulence, ν_T should be constant, so one can test this hypothesis by plotting $\nu_T(k,i) = \nabla.(\bar{U}_k \otimes \bar{U}_k)_i / \Delta\bar{U}_i$. Chollet-Lesieur[1980] found that ν_T has the form

$$\nu_T = 0.267 + 9.21 e^{-3.03\frac{k_c}{k}}.$$

Thus the concept of turbulent diffusion is good for most of the interaction between high modes and low modes as long as the difference of wave vector is large.

9. CONCLUSION

Inhomogeneous non isotropic turbulence is likely to be harder than isotropic homogeneous turbulence and therefore direct simulation of turbulent flow is theoretically out of reach today for high Reynold numbers. However the general behaviour of isotropic turbulence is somewhat understood in the Fourier space. Direct simulation with spectral methods and a "subgrid scale" model allows the prediction of slowly varying turbulences. The method has been extended to simple flows like Poiseuille flows between two parallel flat plates (Moin-Kim[1982], Patera[1984]) and mixing layers (Lesieur[1987]).

CHAPTER 3

REYNOLDS HYPOTHESIS

1. INTRODUCTION

Although direct simulation of turbulent flow may be the way of the future, when computers allow it, the numerical simulation of industrial flows cannot wait until then and something must be done to compute the flow; this is the object of turbulence modeling. It was said earlier that it is not possible to compute the means accurately without modeling the interactions due to the oscillatory parts (eddies...) which fall below the computational grid.

Here we take a different approach, opposed to subgrid scale modeling: *we assume that the flow has two scales so that there is an underlying principle to separate the means from the oscillations* .

In some flows this separation of scales is obvious; for example, behind a cylinder there are the big eddies of the Karman vortex street and the small eddies which come from the boundary layers. However far down stream in the wake of the cylinder there are no such separation of scales. When it is so, the problem must be reformulated: some randomness is introduced in the initial conditions to make the flow random; then the two scales are of a different nature: one set of scales come from the oscillation of the flow; the other set comes from its randomness. Then the means are stochastic averages.

In any case to separate the means from the oscillations we need a *filter.* In subgrid scale modeling (SGS) the filter is attached to the grid of the numerical method. Here the filter is either built into the physics of the flow or, most of the time, comes from the randomness of the initial conditions.

We will first discuss the notion of *filter* and then analyse the filtered Navier-Stokes equations for incompressible fluids. Those filtered equations are not closed (some information has been lost in the filtering process), so the closure hypothesis of *Reynolds* is presented and analysed. The *Smagorinsky* hypothesis is presented also with two *Finite Element Methods* to solve this model numerically.

2. FILTERS

2.1 Definitions

We retain the idea that

$$u = \langle u \rangle + u'$$

where u' is the non computable part or the non relevant part and $\langle u \rangle$ is the mean part. To give form to this statement we must have a filter $\langle \cdot \rangle$. For example, if $\{F(u)_k\}$ denotes the discrete Fourier transform of u

$$F(u)_k(t) = (\frac{1}{2\pi})^3 \int_{R^3} u(x,t)e^{-ik.x}dx,$$

and π_N the truncation operator which replaces $\sum_{k_i=0,1,\dots}$ by $\sum_{k_i=0,1,\dots,N}$ then

$$\langle u \rangle_F = F^{-1}\pi_N F(u) = \sum_{|k|\leq N} F(u)_k e^{ik.x}$$

is a *low pass filter or Fourier filter* . Alternatively it is also

$$\langle u_i \rangle_F = \int_{R^3} u_i(x-y)\frac{\sin(Ny_i)}{Ny_i}dy \equiv u_i * (\frac{\sin Ny_i}{Ny_i})$$

We have seen that for the Navier-Stokes equations with random initial data the natural filter is the *statistical average* ,

$$\langle u \rangle_E = \int u(x,t,\omega)d\omega.$$

Other important filters include

$$\langle u \rangle_B = \frac{1}{|B|}\int_{B(x,r)} u(y,t)dy \quad \text{space averaging}$$

$$\overline{u} = \frac{1}{T_1}\int_{t-T_1}^{t} u(x,\tau)d\tau \quad \text{time averaging}$$

$$\langle\langle \overline{u} \rangle\rangle = \frac{1}{T_1}\int_{t-T_1}^{t} \langle u \rangle_B d\tau \quad \text{space-time averaging}$$

where $B(x,r)$ is the ball of center x and radius r.

2.2 Properties

i) Filters are usually *linear operators* :

$$\langle u + \lambda v\rangle = \langle u\rangle + \lambda\langle v\rangle, \quad \forall u,v \in L^2(R^3\times]0,T'[), \quad \forall \lambda \in R.$$

ii) For some filters *derivatives and averages commute* . For example with the low pass filter:

$$\langle \partial_x u_i\rangle_F = \int_{R^3} \frac{\sin(Ny_i)}{Ny_i} \partial_z u_i|_{z=x-y} dy$$

$$= \partial_x \int_{R^3} u_i(x-y) \frac{\sin(Ny_i)}{Ny_i} dy = \partial_x \langle u\rangle_F.$$

The time averaging filter commutes with spatial and time derivation

$$\overline{\partial_x u} = \frac{1}{T_1}\int_{t-T_1}^{t} \partial_x u(x,\tau)d\tau = \partial_x \frac{1}{T_1}\int_{t-T_1}^{t} u(x,\tau)d\tau$$

$$\overline{\partial_t u} = \frac{1}{T_1}\int_{t-T_1}^{t} \partial_t u(x,\tau)d\tau = \frac{1}{T_1}[u(x,t) - u(x,t-T_1)]$$

$$\partial_t \overline{u} = \partial_t \frac{1}{T_1}\int_{t-T_1}^{t} u(x,\tau)d\tau = \frac{1}{T_1}[u(x,t) - u(x,t-T_1)]$$

Similarly the space average satisfies also this property.

iii) *Double averages* . Filters should have no effect on filtered variables: $\langle\langle u\rangle\rangle = \langle u\rangle$.

For example

$$\langle\langle u\rangle_F\rangle_F = F^{-1}\pi_N F F^{-1}\pi_N F(u) = F^{-1}\pi_N F(u) = \langle u\rangle_F.$$

Notice that the spatial averaging operator does not satisfy this property; in particular

$$\langle\langle u\rangle\rangle = \frac{1}{|B|^2}\int_{B(x,r)}\int_{B(z,r)} u(y,t)dydz \neq \frac{1}{|B|}\int_{B(x,r)} u(y)dy,$$

neither does the time averaging filter unless u is $T_1 - $*periodic*.

Important Remark
If $\langle\langle u\rangle\rangle \neq \langle u\rangle$ then $u = \langle u\rangle + u'$ does not imply that $\langle u'\rangle = 0$. Consequently $\langle u\rangle$ has an oscillatory part still and $u - \langle u\rangle$ has a mean part; it is not really a filter.

iv) *Product Average.* More generally we may require that

$$\langle v\langle u\rangle\rangle = \langle v\rangle\langle u\rangle$$

The statistical average satisfies this property but the other filters do not.

Proposition
The Fourier filter and the statistical filter satisfy all three properties (i),(ii),(iii). Only the statistical filter satisfies property (iv).

3. REYNOLDS STRESS

3.1 The Problem

Let u be a (random) solution of

$$\partial_t u + \nabla.(u \otimes u) + \nabla p - \nu\Delta u = 0, \quad \nabla.u = 0 \text{ in } \Omega\times]0,T'[,$$
$$u(x,0) = u^0(x,\omega), \quad u|_\Gamma = u_\Gamma,$$

where $u(x,.)$ is random and vector valued in R^3. Let $\langle\cdot\rangle$ be the statistical average operator with respect to the probability law induced by u^0. Can we calculate $\langle u\rangle$, $\langle u \otimes u\rangle$, $\nu\langle|\nabla u|^2\rangle$...?

3.2 Reynolds computation

With a filter which satisfies the four properties of §2, denote

$$U = \langle u\rangle \quad u' = U - \langle u\rangle,$$

The filter is applied to the divergence equation:

$$0 = \langle\nabla.u\rangle = \nabla.\langle u\rangle = \nabla.U.$$

Similarly

$$\begin{aligned} 0 &= \langle\partial_t u\rangle + \langle\nabla.(u\otimes u)\rangle + \langle\nabla p\rangle - \nu\langle\Delta u\rangle \\ &= \partial_t\langle u\rangle + \nabla.\langle u\otimes u\rangle + \nabla\langle p\rangle - \nu\Delta\langle u\rangle \\ &= \partial_t U + \nabla.(U\otimes U) + \nabla.[\langle u\otimes u\rangle - U\otimes U] + \nabla P - \nu\Delta U \end{aligned}$$

But

$$\begin{aligned}\langle u \otimes u\rangle &= \langle (U+u') \otimes (U+u')\rangle \\ &= \langle U \otimes U\rangle + \langle U \otimes u'\rangle + \langle u' \otimes U\rangle + \langle u' \otimes u'\rangle \\ &= U \otimes U + \langle u' \otimes u'\rangle.\end{aligned}$$

Theorem
When the filter satisfies all four properties of §2.2, the mean velocity $U = \langle u\rangle$ satisfies the so called Reynolds equations:

$$\partial_t U + U\nabla U + \nabla P - \nu\Delta U + \nabla.\langle u' \otimes u'\rangle = 0,$$
$$\nabla.U = 0,$$

$$U_{t=0} = \langle u^0\rangle, \quad U|_\Gamma = \langle u_\Gamma\rangle.$$

The "Reynolds Stress tensor" is defined by

$$R = -\langle u' \otimes u'\rangle$$

Remark
Note that u' verifies

$$\partial_t u' + u'\nabla U + U\nabla u' + \nabla p' - \nu\Delta u' - \nabla.\langle u' \otimes u'\rangle + \nabla.(u' \otimes u') = 0,$$

$$\nabla.u' = 0.$$

Remark
When the filter satisfies only the first three properties of §2.2 (i)-(iii), it is still possible to define a Reynolds stress tensor. For example, with the time average operator the same computation can be carried out but the derivation is slightly different for the momentum equation. It is filtered as before and since the time-filter commutes with spatial and time derivatives we have:

$$0 = \partial_t\overline{u} + \nabla.(\overline{u \otimes u}) + \nabla\overline{p} - \nu\Delta\overline{u}.$$

It can be written as a Reynolds equation

$$= \partial_t\overline{u} + \nabla.(\overline{u} \otimes \overline{u}) + \nabla\overline{p} - \nu\Delta\overline{u} - \nabla.R$$

But now $R = -\overline{u \otimes u} + \overline{u} \otimes \overline{u}$.

4. REYNOLDS CLOSURE HYPOTHESIS

Experiments suggested to Reynolds that $R = -\langle u' \otimes u' \rangle$ is correlated to ∇U, or for reasons of symmetry to $\nabla U + \nabla U^T$:

$$R = R(\nabla U + \nabla U^T).$$

This assumption is quite reasonable, because turbulences are often in zones of strong gradients of the flow. However there is an obvious counter example to this hypothesis:

• Near the leading edge of a wing the flow is laminar even though the gradients are strong.

5. FRAME INVARIANCE

We proceed to show that $R(\nabla u + \nabla u^T)$ cannot be arbitrarily chosen. The proof relies on frame invariance, a property that every turbulence model user would want: the model should yield results independent of the frame of reference chosen to do the simulation. This type of argument is used thoroughly in rheology. It has been used on fluids by Chorin et al [1979], Chacon et al [1986], Speziale[1988][1988].

5.1 Translation invariance

Suppose $x = y + Z$ where Z is a constant vector of R^3. Then

$$\frac{dx}{dt} = u(x,t) = \frac{dy}{dt}, \quad \frac{\partial u}{\partial x_i} = \frac{\partial u}{\partial y_i}.$$

In (x,t) the velocity is $u(x,t)$ and it verifies:

$$\partial_t u + u\nabla u + \nabla p - \nu\Delta u = 0, \quad \nabla . u = 0.$$

In (y,t) the velocity is $v(y,t) = u(x+Z,t)$ and hence it verifies

$$0 = [\partial_t u + u\nabla u + \nabla p - \nu\Delta u]_{x=y+Z} = \partial_t v + v\nabla_y v + \nabla_y p - \nu\Delta_y v$$

$$\nabla . u|_{x=y+Z} = \nabla_y . v = 0.$$

Therefore the Navier-Stokes equations are invariant with respect to translation. Reynolds equations should also be invariant under translation, otherwise the results of the model would depend upon the choice of a reference frame.

But notice that $\nabla_y v = \nabla_x u$; so $\nabla_y.R(\nabla_y v + \nabla_y v^T) = \nabla_x.R(\nabla_x u + \nabla_x u^T)$ and hence

$$[\partial_t u + u\nabla u + \nabla p - \nu\Delta u - \nabla_x.R(\nabla_x u + \nabla_x u^T)]_{x=y+Z}$$
$$= \partial_t v + v\nabla_y v + \nabla_y p - \nu\Delta_y v - \nabla_y.R(\nabla_y v + \nabla_y v^T).$$

So Reynolds equations are also invariant under translation.

5.2 Galilean Invariance

Suppose now that $x = y + wt$ where w is a constant vector. Then

$$\frac{dx}{dt} = u(x,t) = u(y + wt, t) = \frac{dy}{dt} + w = v + w$$

so let $v(y,t) = u(y+wt,t) - w$ and Navier-Stokes equations for v become

$$\nabla_y.v = \nabla_x.(u - w) = \nabla.u = 0,$$

and, noticing first that $\partial_t v = \partial_t(u - w) + w\nabla_x u = \partial_t u + w\nabla_x u$,

$$\partial_t v + v\nabla_y v + \nabla_y p - \nu\Delta_y v$$
$$= \partial_t u + w\nabla_x u + (u - w)\nabla_x(u - w) + \nabla_x p - \nu\Delta_x(u - w)$$
$$= \partial_t u + u\nabla u + \nabla p - \nu\Delta u = 0.$$

Therefore the Navier-Stokes equations are Galilean invariant. Moreover $\nabla_x = \nabla_y$ so $\nabla_y.R(\nabla_y v + \nabla_y v^T) = \nabla_x.R(\nabla_x u + \nabla_x u^T)$, therefore, Reynolds equations are also Galilean invariant.

5.3 Rotation invariance

Let M be a rotation matrix, that is, which verifies

$$(M^T M)_{kj} = M_{ik}M_{ij} = (MM^T)_{kj} = M_{ki}M_{ji} = \delta_{kj}.$$

Let $x = My$ then, if v denotes the velocity in the y variable,

$$u = \frac{dx}{dt} = M\frac{dy}{dt} = Mv;$$

also for any f we have

$$\frac{\partial f}{\partial x_i} = \frac{\partial f}{\partial y_j}\frac{\partial y_j}{\partial x_i} = \frac{\partial f}{\partial y_j}M^T_{ji}\text{ , that is : } \nabla_x f = M\nabla_y f.$$

Consequently,

$$\nabla u = (M\nabla_y)Mv = M\nabla_y v M^T.$$

Hence

$$\nabla.u = (M\nabla_y).Mv = M_{ij}\frac{\partial}{\partial y_j}M_{ik}v_k = \nabla_y.v,$$

and

$$\begin{aligned}\partial_t u + u\nabla u + \nabla p - \nu\Delta u \\ &= M\partial_t v + (Mv).[(M\nabla_y)Mv] + M\nabla_y p - \nu(M\nabla_y).(M\nabla_y[Mv]) \\ &= M(\partial_t v + v\nabla_y v + \nabla_y p - \nu\Delta_y v).\end{aligned}$$

So the Navier-Stokes equations are rotation invariant. Notice however that if we denote symbolically the momentum equation by ns(u), it has become Mns(v).

Now it follows from the formulae above that

$$\nabla_x u + \nabla_x u^T = M(\nabla_y v + \nabla_y v^T)M^T.$$

This allows us to evaluate the Reynolds stress in both frames of reference:

$$\begin{aligned}\nabla_x.R(\nabla_x u + \nabla_x u^T) &= \nabla_y.[M^T R(M[\nabla_y v + \nabla_y v^T]M^T)] \\ &= M\nabla_y.[M^T R(M[\nabla_y v + \nabla_y v^T]M^T)M]\end{aligned}$$

Let u be a solution of Reynolds' equations. Then v will satisfy the same equations in the y-frame if and only if

$$M^T R(M[\nabla_y v + \nabla_y v^T]M^T)M = R(\nabla_y v + \nabla_y v^T)$$

for all v with $\nabla_y.v = 0$ and all M with $M^{-1} = M^T$. Since all matrices A with zero trace are spanned by $\nabla_y v + \nabla_y v^T$ we must ask that

$$M^T R(MAM^T)M = R(A) \quad \forall A, M \text{ with } \quad M^{-1} = M^T, \qquad trA = 0.$$

5.4 Time dependent rotation invariance

Let $x = M(t)y = \Omega(t) \times y$ where M is a rotation of vector $\Omega(t)$. Then $v = dy/dt$ satisfies the Navier-Stokes equations in a rotating frame, i.e. (see for example Bachelor[1970,p141]):

$$\nabla_y.v = 0,$$

$$\partial_t v + v\nabla_y v + \nabla_y q - \nu\Delta_y v + \partial_t\Omega \times y + \Omega \times (\Omega \times y) + 2\Omega \times v = 0.$$

If $\{u, p\}$ denotes the solution of the Navier-Stokes equations in the x-frame and $\{v, q\}$ the same in the y-frame, they are related by the following formulae:

$$q = p + \frac{1}{2}(\Omega^2 y^2 - (\Omega.y)^2), \quad u = Mv + \partial_t My = \Omega \times v + \partial_t\Omega \times y.$$

Let us see what is required of the Reynolds stress tensor to have a similar invariance for Reynolds equations. Since Reynolds' equations are Navier-Stokes' plus the Reynolds stress term, R, let us compute the effect of such a change of coordinates on R.
For this purpose notice that

$$\nabla u = (M\nabla_y)Mv + \nabla((\partial_t M)M^T x) \quad = M\nabla_y v M^T + M(\partial_t M^T).$$

$$\nabla u + \nabla u^T = M(\nabla_y v + \nabla_y v^T)M^T + M\partial_t M^T + \partial_t M M^T,$$

because $(M\partial_t M^T)^T = (\partial_t M)M^T$. But $MM^T = I$ implies $(\partial_t M)M^T + M\partial_t M^T = 0$ so there are no new property required for R. (See also Speziale[1988]).

Proposition
To be Frame Invariant, the only possible form for a symmetric matrix R, function of another symmetric matrix $A \in R^{d\times d}$, is

$$R(A) = a_0 I + a_1 A + ... + a_{d-1}A^{d-1},$$

where the a_i *are functions of the invariants of* A *only .*
Proof
i) Invariants of A
In two dimensions ($d = 2$), the invariants of a 2×2 matrix A are $tr(A), det(A)$ or $|A| = \sqrt{tr(AA^T)}$; in three dimensions, there is also $|A^2|$. Indeed

$$tr(MAM^T) = M_{ij}A_{jk}M_{ik} = \delta_{jk}A_{jk} = tr(A),$$

$$det(MAM^T) = dct(M)^2 dct(A) = det(A),$$

$$|(MAM^T)^2| = |MAM^T MAM^T| = |MA^2M^T| = |A^2|.$$

Now $R = a_0 I + a_1 A + ... + a_{d-1}A^{d-1}$ is frame invariant because the a_i are so and because

$$M^T R(MAM^T)M = M^T(a_0 I + a_1 MAM^T + ... + a_{d-1}(MAM^T)^{d-1})M$$

$= a_0$ I $+ a_1 A$ +...+ $a_{d-1}A^{d-1}$.

ii) Decomposition of A^n on eigen matrices

Conversely, following Rivlin-Eriksen (cf Ciarlet[1986]), assume first that $d = 3$ and that A has 3 distinct eigen values and let $\{\lambda_i, p_i\}$ be a set of eigen values and orthonormal eigen vectors (recall that A is symmetric). Since $Span\{I, A, A^2\} = Span\{p_i p_i^T\}_{i=1,2,3}$ we have necessarily

$$I = \sum_{i=1,2,3} p_i p_i^T; \quad A = \sum_{i=1,2,3} \lambda_i p_i p_i^T; \quad A^2 = \sum_{i=1,2,3} \lambda_i^2 p_i p_i^T.$$

because any of these identity gives the right answer when a dot-product with p_j is made. For example:

$$(I - \sum_{i=1,2,3} p_i p_i^T)p_j = p_j - \sum_{i=1,2,3} p_i(p_i^T p_j) = p_j - \sum_{i=1,2,3} p_i \delta_{ij} = 0.$$

We shall show below that $R(A)$ is diagonalizable in the same basis as the $p_i p_i^T$. This means that for some μ_i and some a, b, c, the following will be true:

$$R(A) = \sum \mu_i p_i p_i^T \quad \Leftrightarrow \quad R(A) = aI + bA + cA^2,$$

where a,b,c depend only on $\{\lambda_i\}$ because they are found by inverting the linear system for the $p_i p_i^T$.

iii) A and R are diagonalizabled in the same basis

Let Q^i be the diagonal matrix with -1 on the diagonal except at the entry i where there is +1, i.e.

$$Q^i_{jk} = \pm\delta_{jk}, \quad Q^i_{jj} = -1 \text{ if } j \neq i\ ,\ Q^i_{ii} = 1$$

It is easy to check that MQ^i means replacing the i-th column of M by its opposite in sign. Notice also that Q is a rotation matrix. Hence if M is a matrix of eigen vectors of A, so is MQ, therefore

$$(MQ)^T A(MQ) = \Lambda = M^T AM,$$

where $\Lambda =$ diag $(\lambda_1, \lambda_2, \lambda_3)$.

Now frame invariance requires that

$$Q^T M^T R(A) M Q = Q^T R(M^T A M) Q = R(Q^T M^T A M Q)$$
$$= R(M^T A M) = M^T R(A) M.$$

Hence $C = M^T R(A) M$ satisfies

$$Q^{iT} C Q^i = C, \quad i = 1, 2, 3.$$

It is not hard to see that this implies that C is diagonal. Let us do it in two dimensions:

$$\begin{pmatrix} a & b \\ b & c \end{pmatrix} = \begin{pmatrix} 1 & 0 \\ 0 & -1 \end{pmatrix} \begin{pmatrix} a & b \\ b & c \end{pmatrix} \begin{pmatrix} 1 & 0 \\ 0 & -1 \end{pmatrix} = \begin{pmatrix} a & -b \\ -b & c \end{pmatrix}$$

For the rest of the proof, i.e. if A has only two or one distinct eigen value see Ciarlet[1986].

Corollary

In two dimensions, frame invariance and the assumption that Reynolds' tensor be a function of $\nabla u + \nabla u^T$ only, imply that

$$R(\nabla u + \nabla u^T) = a(|\nabla u + \nabla u^T|) I + \mu(|\nabla u + \nabla u^T|)(\nabla u + \nabla u^T),$$

giving for Reynolds' equations

$$\partial_t u + u \nabla u + \nabla p - \nu \Delta u - \nabla.[\mu(|\nabla u + \nabla u^T|)(\nabla u + \nabla u^T)] = 0,$$
$$\nabla.u = 0.$$

In three dimensions the same imply

$$R(\nabla u + \nabla u^T) = aI + \mu(\nabla u + \nabla u^T) + \lambda(\nabla u + \nabla u^T)^2$$

where a, μ, λ are functions of $|\nabla u + \nabla u^T|$ and $|(\nabla u + \nabla u^T)^2|$ only. Then Reynolds' equations take the form

$$\partial_t u + u \nabla u + \nabla p - \nu \Delta u - \nabla.[\mu(\nabla u + \nabla u^T) + \lambda(\nabla u + \nabla u^T)^2] = 0,$$
$$\nabla.u = 0.$$

Remark

As $\nu \Delta u = \nu \nabla.(\nabla u + \nabla u^T)$, this molecular viscosity term can be removed by changing μ into $\mu + \nu$.

6. ALGEBRAIC SUBGRID-SCALE MODELS

6.1 Smagorinsky's model

Smagorinsky proposed

$$\mu = ch^2|\nabla u + \nabla u^T|,$$

that is

$$R = ch^2|\nabla u + \nabla u^T|(\nabla u + \nabla u^T) + aI, \quad c \cong 0.01$$

where $h(x)$ is the mesh size of the numerical method around point x. Changing the pressure p into $p - a$ amounts to set $a = 0$, so it is not essential to know a.

This hypothesis is compatible with the symmetry and a bidimensional analysis of R ; it is reasonable in 2D but not sufficient in 3D (see also Baker [1985]).

The fact that h is involved is justified by ergodicity [1]. Recall that the ergodic theorem says that statistical averages are equivalent to time or ensemble averages. By using an ensemble average over spheres of radius h as filter in the definition of Reynolds' stress, it makes sense to assume that R depends upon h. The correct power of h is found by dimensional analysis.

Numerical experiments show satisfactory results (Moin-Kim [1982]) when there are enough points. This model performs well when the small scale, which is the mesh size, is in Kolmogorov's inertial range. In other words, to check that enough points have been used one may proceed as follows:

- Choose a mesh size and solve the problem numerically.
- Fourier transform the velocity and plot $|u|$ versus K.
- Check that the end of the energy spectrum behaves like $K^{-5/3}$. If not refine the mesh.

Remark

Mathematically the Smagorinsky system is better than Navier-Stokes' because there is existence, uniqueness and regularity even in three dimensions (Lions[1968]).

6.2 Métais-Lesieur's model

[1] Ergodicity for Navier-Stokes equations with random initial data is an open problem.

Starting from the spectral approximation of ν_T given in §II.8 and searching for an equivalent formula in the physical space, Métais-Lesieur[1992] and David[1993] proposed the following

$$\mu = 0.104\Phi(\alpha)h^2|\nabla u|$$

where Φ is 1 if $\alpha > 20^o$ and zero otherwise; α is the angle between $\nabla \times u(x)$ and its local mean value computed by averaging the values of $\nabla \times u$ on a small domain around x.

7. FINITE ELEMENT METHODS FOR SMAGORINSKY'S MODEL

Consider the Navier-Stokes equations with non constant viscosity $\nu(x,t) > \nu_0 > 0$ and homogeneous boundary data

$$\partial_t u - \nabla.(\nu[\nabla u + \nabla u^T]) + u.\nabla u + \nabla p = f \ ,$$
$$\nabla.u = 0 \qquad in\ \Omega\times]0,T'[,$$
$$u|_{t=0} = u^0 \qquad \text{in } \Omega \quad u = 0 \text{ on } \Gamma\times]0,T'[\quad (\Gamma = \partial\Omega).$$

Denote

$$J_0(\Omega) = \{v \in H_0^1(\Omega)^d : \nabla.v = 0\}.$$

A weak formulation in $J_0(\Omega)$ is

$$(\partial_t u, v) + (\frac{\nu}{2}(\nabla u + \nabla u^T), \nabla v + \nabla v^T) + (u\nabla u, v) = (f, v) \qquad \forall v \in J_0(\Omega)$$

A spatial approximation of $J_0(\Omega)$ can be constructed from finite element approximations $V_{0h} \approx H_0^1(\Omega)^d$ and $Q_h \approx L^2(\Omega)$ as follows.

7.1 Spatial approximation

Find u_h such that

$$u_h \in J_{0h} = \{u_h \in V_{0h} : (q_h, \nabla.u_h) = 0 \quad \forall q_h \in Q_h\}$$
$$(\partial_t u_h, v_h) + (\frac{\nu}{2}[\nabla u_h + \nabla u_h^T], \nabla v_h + \nabla v_h^T) + (u_h.\nabla u_h, v_h)$$
$$= (f, v_h) \qquad \forall v_h \in J_{0h}.$$

It has been shown (Brezzi[1974], Babuska[1978]...) that the method will converge if

$$\inf_{q_h \in Q_h} \sup_{v_h \in V_{0h}} \frac{(\nabla . v_h, q_h)}{|\nabla v_h|_0 |q_h|_0} > \beta.$$

There are many constructions which verify this inequality. For example, given a triangulation T_h of Ω, the space of continuous piecewise quadratic velocities and continuous piecewise linear pressures works:

The P^2/P^1 element

$$V_h = \{v \in C^0(\overline{\Omega})^3 : v_i|_T \in P^2, i = 1,2,3 \quad \forall T \in T_h\},$$
$$Q_h = \{q \in C^0(\overline{\Omega}) : q|_T \in P^1, \quad \forall T \in T_h\};$$

V_{oh} and Q_{0h} are the functions of V_h and Q_h which are zero on Γ. Here P^k denotes the space of polynomials of degree less than k.
The system is a set of ordinary differential equations in time for the values of u_h on the vertices and mid edges. The computational work can be reduced slightly by using the so called P^1 iso P^2/P^1 element which is the same as above but with $v_i|_T \in P^1$ on the mesh obtained by dividing each triangle of V_h into 4 sub triangles joining the midsides.

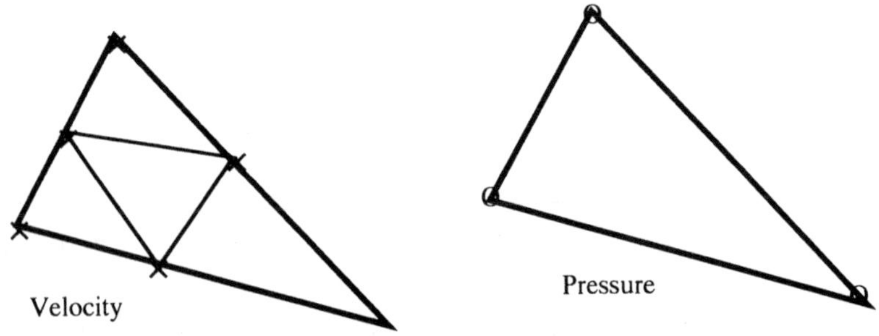

Figure 3.1 : The P^1-iso-P^2/P^1 and the P^2/P^1 element have their degrees of freedom at the vertices for the pressure and at the vertices and mid-sides for the velocity.

Let us give two other examples of spaces of approximation

The $Q^2 - P^1$ element
Let R_h be a quadrangulation of Ω. By dividing each quadrangle in two triangles by a diagonal we associate to R_h a triangulation called T_h. Then choose

$$V_h = \{v \in C^0(\overline{\Omega})^2 : v_i|_R \in Q^2, i = 1,2,3 \quad \forall R \in R_h\}$$

$$Q_h = \{q : q|_T \in P^1, \quad \forall T \in T_h\}$$

where Q^2 denotes the space of quadratic functions in *each* coordinate variable x_i.

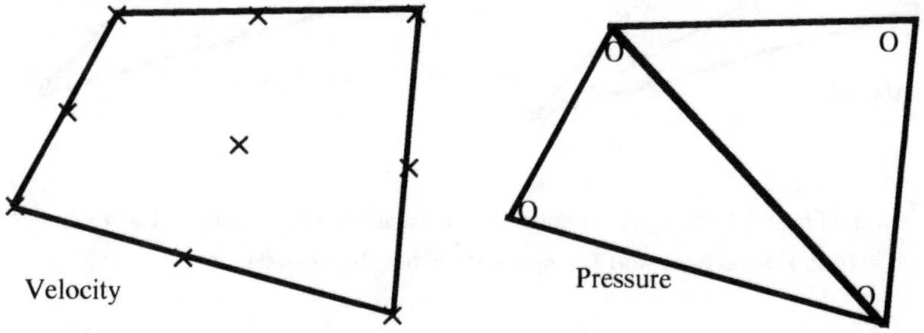

Figure 3.2 :The Q^2/P^1 element has its degrees of freedom at 3 points inside each triangle for the pressure and at the vertices, quadrilateral center and mid-sides for the velocity. The pressure, however, is discontinuous so its degrees of freedom are different on each element.

The P^1 – *bubble*/P^1 element

Let $T_{h'}$ denote the triangulation obtained by dividing all triangles of T_h into 3 sub triangles by joining the vertices to the center. Then choose

$$V_h = \{v \in C^0(\overline{\Omega})^3 : v_i|_T \in P^1, i = 1,2,3 \quad \forall T \in T_{h'}\}$$

$$Q_h = \{q \in C^0(\overline{\Omega}) : q|_T \in P^1, \quad \forall T \in T_h\}$$

Static condensation (elimination) of the degree of freedom at the center of each triangle can be applied to reduce storage (see Pironneau[1989] for example); this element is similar to the one introduced by Hughes et al[1986].

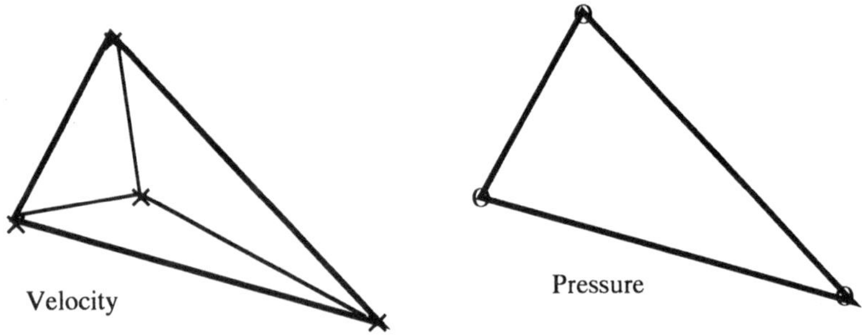

Figure 3.3 :The P^1-bubble/P^1 element has its degrees of freedom at the vertices for the pressure and at the vertices and triangle center for the velocity.

Theorem (Bernardi-Raugel [1985])
Assume that ν is bounded from below by ν_o. With the P1-iso-P2 /P1 or the P1-bubble / P1 element the following error estimates hold.

$$||u - u_h||_{L^2(0,T';H^1)} < \frac{c}{\nu_0} h |u\nabla u|_0,$$

$$||p - p_h||_{H^{-1}(0,T';L^2)} < \frac{c}{\nu_0} h |u\nabla u|_0,$$

$$||u - u_h||_{L^2(]0,T'[\times\Omega)} < \frac{c}{\nu_0} h^2 |u\nabla u|_0,$$

where c is a function of $||u||_{L^4(]0,T'[\times\Omega)}$.

7.2 Consequences
The Kolmogorov hypothesis enables us to estimate $|u\nabla u|_0$. For an isotropic homogeneous turbulence $\langle u\nabla u\rangle \approx f(\epsilon,\nu)$ and by a dimensional analysis $\langle u\nabla u\rangle \approx U^2/L = \epsilon^{3/4}\nu^{-1/4}$. Thus *to compute stresses, vorticities and pressures, a mesh $h << \nu^{5/4}$ is needed. To compute velocities a mesh $h << v^{5/8}$ is sufficient.* This result is a little worse than $h < \nu^{3/4}$ suggested by the Kolmogorov spectrum, except if only the velocities are needed in which case it is a little better.

7.3. Approximation in Time: The Characteristic Galerkin Method
In turbulent flows the convective terms dominate. Thus it is necessary to use upwinding in one form or another. Let us recall two techniques for upwinding.

Let $X(x,t;s)$ be the solution of

$$\frac{dX}{ds} = u(X,s); \quad X|_{s=t} = x$$

$X(x,t;s)$ is the position at time s of the particle which was at position x at time t; X is also the characteristic of Euler's momentum equation. Notice that

$$\partial_t u + u\nabla u \approx \frac{1}{\delta t}(u^{n+1} - u^n o X^n),$$

where $X^n(x) = X(x, t^{n+1}; t^n)$, (roughly, $X^n(x) \approx x - u(x,t^n)\delta t$). Thus we obtain the following scheme :

$$\frac{1}{\delta t}(u_h^{n+1}, v_h) + (\frac{\nu^n}{2}[\nabla u_h^{n+1} + \nabla u_h^{n+1^T}], \nabla v_h + \nabla v_h^T)$$

$$= (f^n, v_h) + \frac{1}{\delta t}(u_h^n o X_h^n, v_h), \quad \forall v_h \in J_{0h}$$

where $X_h^n(x)$ is a numerical approximation of $X^n(x)$.
We note that this is a positive definite linear system, which must be solved at each time step. If X_h^n is well chosen, this scheme is unconditionally stable, and convergent in $0(\delta t + h)$.
In practive the linear system is converted into

$$\begin{pmatrix} A & B \\ B^T & 0 \end{pmatrix} \begin{pmatrix} U \\ P \end{pmatrix} = \begin{pmatrix} F \\ 0 \end{pmatrix}$$

because the space J_{0h} has constraints in its definition; P is the vector of Lagrange multipliers of those constraints. More details can be found in Pironneau[1987].
There are two differences with the Navier-Stokes equations with constant viscosity:
• The matrix of the linear system must be reconstructed at each time step because ν_T depends on time.
• The form $\nabla.(\nu_T[\nabla u + \nabla u^T]$) couples the components u_i of the velocities and A is no longer a bloc matrix.
For these reasons a good method to solve the linear system is the biconjugate gradient method such as ORTHOMIN or ORTHODIR.

Remark
In practice, the exact calculation of the integral $(u_h^n o\ X_h^n, v_h)$ is unnecessarily costly; a direct quadrature Gauss formula is used:

$$(u_h^n o X_h, v_h) \cong \sum_{i \in I} u_h(X_h^n(\xi^i)) v_h(\xi^i) \pi_i$$

or a dual formula (Benqué et al. [1980])

$$(u_h^n o X_h, v_h) \cong \sum_{i\in I} u_h(\xi^i) v_h(X_h^{n^{-1}}(\xi^i))\pi_i.$$

Here I is the set of quadrature points ξ_i and π_ι are the weights in the quadrature formula.

7.4 Approximation in time: a SUPG formulation

A *Petrov-Galerkin* variational formulation for the Navier-Stokes is used with a small parameter τ and the test functions equal to $v_h + \tau u_h \nabla v_h$ instead of v_h :

$$(\partial_t u_h + u_h \nabla u_h + \nabla p_h, v_h + \tau u_h \nabla v_h) + (\frac{\nu}{2}[\nabla u_h + \nabla u_h^T], \nabla v_h + \nabla v_h^T)$$

$$-(\tau u_h \nabla v_h, \nabla . \frac{\nu}{2}[\nabla u_h + \nabla u_h^T]) = (f, v_h + \tau u_h \nabla v_h), \quad \forall v_h \in V_{0h};$$

$$(\nabla . u_h, q_h) = 0, \quad \forall q_h \in Q_h.$$

The last integral on the left hand side is understood as a sum of integrals on each element of the triangulation because ∇u_h jumps at the elements interfaces.

This is the simplest SUPG method where the viscosity is added in space only. Johnson [1987] rightly suggest in his error analysis to use a space-time discretisation for V_h and Q_h (see VIII.4.3).

A semi-implicit time discretisation gives the following scheme:

$$(\frac{1}{\delta t}[u_h^{n+1} - u_h^n] + u_h^n \nabla u_h^{n+1}, v_h + \tau u_h^n \nabla v_h) + (\nabla p_h^{n+1}, v_h) + (\nabla p_h^n, \tau u_h^n \nabla v_h)$$

$$+(\frac{\nu^n}{2}[\nabla u_h^{n+1} + \nabla u_h^{n+1^T}], \nabla v_h + \nabla v_h^T) - (\tau u_h^n \nabla v_h, \nabla . [\frac{\nu}{2}[\nabla u_h + \nabla u_h^T]])$$

$$= (f^{n+1}, v_h + \tau u_h^n \nabla v_h), \quad \forall v_h \in V_{0h}$$

$$(\nabla . u_h^{n+1}, q_h) = 0, \quad \forall q_h \in Q_h.$$

A good choice of the parameter τ is critical. One possibility is to define a vector $h(x) = \{\delta x_1, \delta x_2\}$ where δx_i is the mesh size in the direction x_i at point x and set

$$\tau = \frac{h.u}{2|u|^2}.$$

For more details the reader is sent to Johnson [1987], Hughes[1987]...

7.5 Wall laws

Consider a boundary condition coming from a wall law, like the one presented in chapter 1:

$$u.n = 0,$$

$$\frac{u.s}{\sqrt{\nu|\frac{\partial u.s}{\partial n}|}} - \frac{1}{0.41}\log(\delta\sqrt{\frac{1}{\nu}|\frac{\partial u.s}{\partial n}|}) + 5.5 = 0.$$

Let us rewrite the second equation symbolicaly as

$$\frac{\partial u.s}{\partial n} = -g(u.s).$$

With these boundary conditions, consider the generalized Stokes problem

$$\alpha u - \nabla.(\nu_T[\nabla u + \nabla u^T]) + \nabla p = f, \quad \nabla.u = 0.$$

By multiplication by v and integration, we obtain, for all v and all q:

$$\alpha(u,v)+(\frac{\nu_T}{2}[\nabla u+\nabla u^T],\nabla v+\nabla v^T)-\int_\Gamma v.[\nabla u+\nabla u^T]n+(\nabla p,v) = (f,v)$$

$$-(u,\nabla q) + \int_\Gamma u.nq = 0.$$

But $v.[\nabla u + \nabla u^T]n = v.s\partial_s u.n + \partial_n u.v = v.s\partial_n u.s$, so a variational formulation is

$$\alpha(u,v) + (\frac{\nu_T}{2}[\nabla u + \nabla u^T], \nabla v + \nabla v^T) + \int_\Gamma g(u.s)v.s + (\nabla p, v)$$

$$= (f,v), \quad \forall v \in H^1(\Omega)^3$$

$$(u,\nabla q) = 0 \quad \forall q \in H^1(\Omega)/R.$$

It shows that if g is positive and monotone there will be a solution (see Pares[1988])

This method, applied to the Smagorinsky model, will give the following Characteristic- Galerkin algorithm

$$\frac{1}{\delta t}(u_h^{n+1}, v_h) + (\frac{\nu^n}{2}[\nabla u_h^{n+1} + \nabla u_h^{n+1^T}], \nabla v_h + \nabla v_h^T)$$

$$+ \int_\Gamma g_1(u_h^n.s) u_h^{n+1}.s v_h.s + (\nabla p_h^{n+1}, v_h) = (f^n, v_h) + \frac{1}{\delta t}(u_h^n o X_h^n, v_h),$$

$$\forall v_h \in V_h, \quad (u_h^{n+1}, \nabla q_h) = 0 \quad \forall q \in Q_h,$$

where $g_1(z) = g(z)/z$, and where V_h, Q_h are the Finite Element spaces described above; they contain no conditions on the boundary; all the boundary conditions have been implemented weakly.

8. NUMERICAL RESULTS

Smagorinsky's model was used by Moin-Kim[1982], in a Finite Difference Navier-Stokes solver, to simulate flows over a flat plate.
They report excellent agreement with experiments but they also use a lot of points.
With Finite elements, an implementation of this method was done at INRIA by F. Hecht and C. Parès[1988] in three dimensions. It uses Characteristic-Galerkin upwinding with the $P^1 - bubble/P^1$ element with tetrahedra.
The method was tested on an automobile to compute its drag coefficient. From the turbulence point of view (figure 3.4) the results reported here are not very good because there are too few grid-points. However from the point of view of large eddy simulation (figure 3.5) the results are reasonably good; the model has a stabilizing effect on the flow and the position of the main eddies agrees with experiments. Better results can be obtained with Smagorinsky's model with finer meshes.

9. CONCLUSION

To filter the Navier-Stokes equations is an interesting concept but it requires a closure hypothesis.
With only one such hypothesis, that the Reynolds tensor be a local function of ∇u, we have shown that it is possible reduce the unknown functions to one scalar function in two dimensions and two unkown functions of two unknowns in three dimensions.
The subgrid scale turbulence model of Smagorinsky is one such simple model where the unknown function is assumed linear.
However Smagorinsky's hypothesis is too simple for many applications. In the next chapter we will discuss the $k - \epsilon$ model which retains a

slightly more general form for Reynolds hypothesis by assuming that Reynolds' tensor is a function of ∇u but also of the kinetic energy k of the small eddies and of ϵ the rate of dissipated energy by the small eddies.

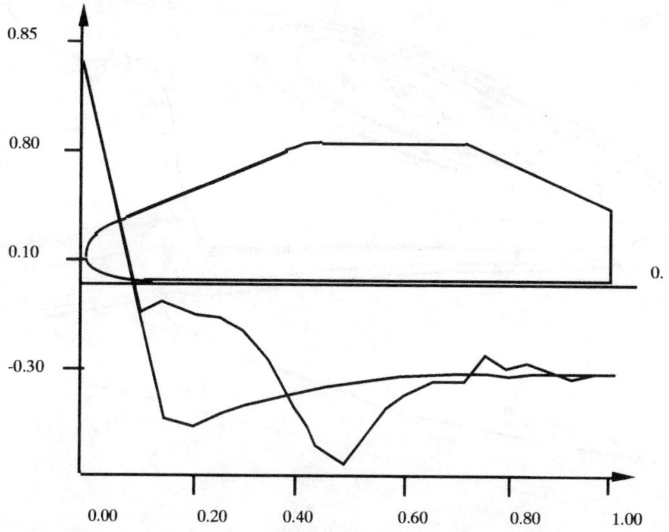

Figure 3.4 Pressure distribution on the symmetry plane of the car computed by Smagorinsky's model: notice that the peak of pressure on the roof is unphysical. This is because the grid is too coarse (8000 vertices). This figure illustrates the stabilizing effect of the Smagorinsky model but also its inefficiency when the mesh is too coarse. (Courtesy of F. Hecht[1989])

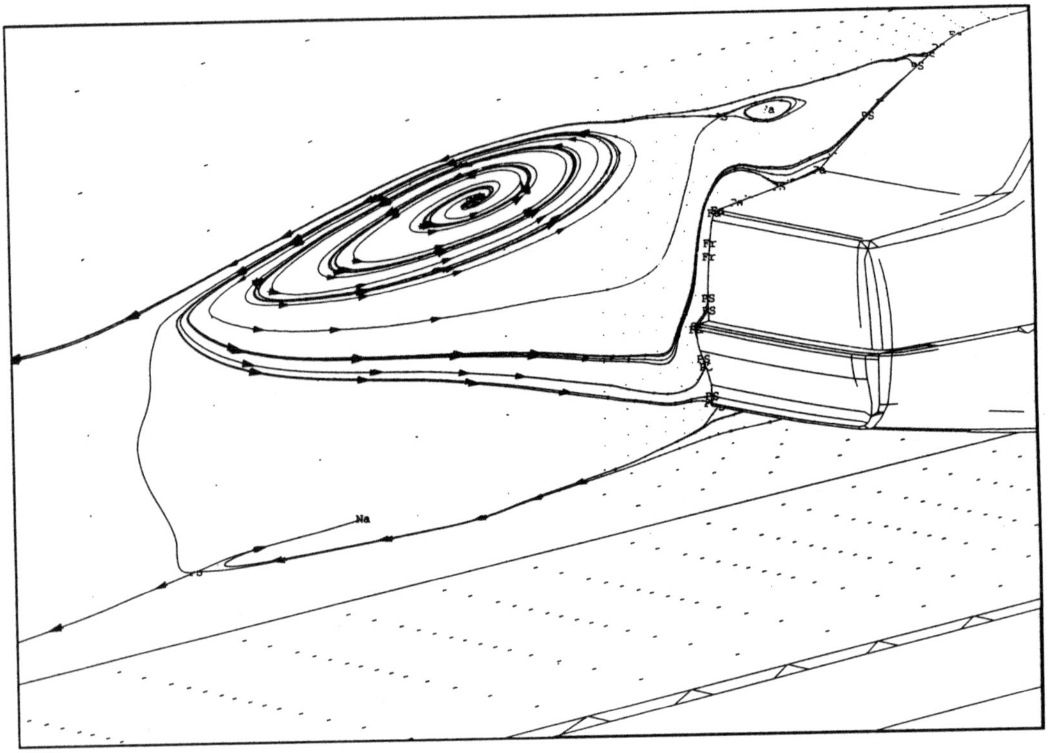

Figure 3.5 Particle path and pressure map behind a car; the position of the main eddies is correct. The grid has 25 000 vertices (Courtesy of F. Hecht[1989])

CHAPTER 4

THE K-EPSILON MODEL

1. INTRODUCTION

Smagorinsky's model is too simple and requires too many discretization points to perform well. For boundary layers, Prandtl [1945] suggested a partial differential equation for $l = \nu_T/u^*$ of the type

$$\partial_t l + u\nabla l + l|\nabla u + \nabla u^T|^2 + ... = 0.$$

Rotta [1951] improved the model and suggested a two-equations model for two turbulent scales rather than one. But the most widely used two-equations turbulence model was introduced by Launder and Spalding [1972] , the so called $k-\epsilon$ model . It consists of two equations for the turbulent kinetic energy k and for ϵ, the rate of dissipation of the turbulent energy. The model was latter extended by Patankar [1980], Rodi [1972],... to take into account different complex situations including curvature, non isotropic turbulence, buoyancy effects and so on. There are several attempts to do better in particular with the so called *Reynolds stress* models. In these models, a new transport-diffusion equation is introduced for each Reynolds tensor component [1975]. But while more complex by an order of magnitude, it is not clear that they perform an order of magnitude better. Despite the fact that the validity of $k-\epsilon$ is not universal, it presents a good compromise between simplicity and generality.

This chapter is dedicated to the derivation of the $k-\epsilon$ equations from the Navier-Stokes equations. We present first the governing equations. We then describe the derivation of the model and the hypotheses necessary to its derivation. Also, we shall justify some of them heuristically. Finally we shall discuss the problem of boundary conditions.

2. DEFINITION OF THE MODEL

2.1 Definition

Given a filter $\langle\cdot\rangle$, the incompressible Reynolds averaged Navier-Stokes equations for the mean flow U and mean pressure P are:

$$\partial_t U + U\nabla U + \nabla P - \nu\Delta U - \nabla.R(k,\epsilon,\nabla U+\nabla U^T) = 0, \quad \nabla.U = 0,$$

where $R_{ij} = -\langle u_i u_j\rangle$ is the Reynolds tensor. The kinetic energy of the turbulence k and the rate of dissipation of turbulent energy ϵ are defined by

$$k = \frac{1}{2}\langle|u'|^2\rangle, \qquad \epsilon = \frac{\nu}{2}\langle|\nabla u' + \nabla u'^T|^2\rangle;$$

then R, k, ϵ are modeled in terms of the mean flow U by

$$R = -\frac{2}{3}kI + (\nu + c_\mu\frac{k^2}{\epsilon})(\nabla U + \nabla U^T),$$

$$\partial_t k + U\nabla k - \frac{c_\mu}{2}\frac{k^2}{\epsilon}|\nabla U + \nabla U^T|^2 - \nabla.(c_\mu\frac{k^2}{\epsilon}\nabla k) + \epsilon = 0,$$

$$\partial_t\epsilon + U\nabla\epsilon - \frac{c_1}{2}k|\nabla U + \nabla U^T|^2 - \nabla.(c_\epsilon\frac{k^2}{\epsilon}\nabla\epsilon) + c_2\frac{\epsilon^2}{k} = 0,$$

with $c_\mu = 0.09$, $c_1 = 0.126$, $c_2 = 1.92$, $c_\epsilon = 0.07$.

2.2 Notations

$$\text{Let } \nu_T = c_\mu\frac{k^2}{\epsilon}, \quad D_t = \frac{\partial}{\partial t} + u\nabla \quad E = \frac{1}{2}|\nabla u + \nabla u^T|^2, P^* = P + \frac{2}{3}k;$$

then the $k-\epsilon$ model and the Navier-Stokes equations take the following form

$$D_t k - \nabla.(\nu_T\nabla k) - c_\mu\frac{k^2}{\epsilon}E + \epsilon = 0,$$

$$D_t\epsilon - \nabla.(\frac{c_\epsilon}{c_\mu}\nu_T\nabla\epsilon) - c_1 kE + c_2\frac{\epsilon^2}{k} = 0,$$

$$D_t U + \nabla P^* - \nabla.[(\nu + \nu_T)(\nabla U + \nabla U^T)] = 0, \quad \nabla.U = 0.$$

3. JUSTIFICATION OF THE MODEL

3.1 Summary of the hypotheses

In order to justify the model heuristically we need the following

H1: Reynolds hypothesis for $\langle u' \otimes u' \rangle$ (this tensor is a function of $\nabla U + \nabla U^T$, k and ϵ only).

H2: Convection by random fields produces diffusion for the mean.

H3: Ergodicity to replace means by spatial averages when necessary.

H4: Isotropy of u' to neglect all odd boundary integrals after integrations by parts.

H5: Reynolds hypothesis for $\langle \omega' \otimes \omega' \rangle$ (this tensor is function of $\nabla U + \nabla U^T$, k and ϵ only).

H6: Quasi Gaussian turbulence so as to neglect $\langle (\omega' \otimes \omega') : \nabla u' \rangle$.

H7: A closure hypothesis to model $\nu^2 \langle |\nabla \omega'|^2 \rangle$ by $c_2\, \epsilon^2/k$.

H8: The coefficient of proportionality between $\langle u' \otimes u' \rangle$ and $\nabla U + \nabla U^T$ is $\nu_T = c_\mu k^2/\epsilon$.

Perhaps the last three are the most questionable of all, and so, as we shall see, the equation for ϵ is less reliable than the one for k.

Notice that k^2/ϵ has the dimension of a viscosity (L^2/T). So hypothesis (H8) on R is compatible with Reynolds hypothesis (hypothesis H1) in two dimensions. In 3 dimensions (H8) also says that R is parallel to $\nabla U + \nabla U^T$ which is much stronger.

3.2 Derivation of the Equation for k

To obtain an equation for k, recall the equation for u' which was obtained by subtracting Reynolds equation from the Navier-Stokes equations (see remark in §II.3.2):

$$\partial_t u' + u' \nabla U + (U + u') \nabla u' + \nabla p' - \nu \Delta u' - \nabla . \langle u' \otimes u' \rangle = 0, \quad \nabla . u' = 0.$$

Multiply this equation by u' and apply the filter (here again we assume that it satisfies the four properties §2.2 in Chapter 3). It yields

$$\partial_t \langle \frac{u'^2}{2} \rangle + \langle u' \otimes u' \rangle : \nabla U + \langle (U + u') \nabla \frac{u'^2}{2} \rangle$$
$$+ \nabla . \langle p' u' \rangle - \nu \langle u' \Delta u' \rangle = 0.$$

By definition of k and R we have then

$$\partial_t k - R : \nabla U + \langle (U + u') \nabla \frac{u'^2}{2} \rangle - \nu \langle u' \Delta u' \rangle + \nabla . \langle p' u' \rangle = 0.$$

Now to model the third term we recall that the convection of a passive scalar by a random velocity field as in

$$\partial_t c + (U + u')\nabla c = 0$$

leads to a convection-diffusion equation for the mean C of c when u is "mixing" (see Chapter 11):

$$\partial_t C + U\nabla C - \nabla.(\kappa \nabla C) = 0$$

where κ is a function of the second order moments of u'. Now

$$\partial_t \frac{u'^2}{2} + (U + u')\nabla \frac{u'^2}{2}$$

can be seen as the convective part of an equation for $\langle u'^2/2 \rangle$ with the convection velocity $u = U + u'$ (hypothesis H2). Hence it is likely to give on the average

$$\partial_t \langle \frac{u'^2}{2} \rangle + \langle (U + u')\nabla \frac{u'^2}{2} \rangle \cong \partial_t k + U\nabla k - \nabla.[\mu(R)\nabla k].$$

Of course this is not correct because u'^2 is not independent of u'. In any case if we assume also that the diffusion $\mu(R)$ is the same as in the Reynolds equation (i.e. $\mu(R) = \nu_T$), then the equation for k is now:

$$\partial_t k + U\nabla k - R : \nabla U - \nabla.(\nu_T \nabla k) - \nu\langle u'\Delta u' \rangle + \nabla.\langle p'u' \rangle = 0.$$

The last two terms cannot be expressed analytically in terms of u, k and ϵ. Therefore, they must be modeled.
For this, one uses an ergodicity hypothesis and replaces the filter by a space average on a ball of centre x and radius r, $B(x, r)$ (hypothesis H3).
Thus $\langle \cdot \rangle$ is replaced by an integral on B

$$\begin{aligned}\langle u'\Delta u' \rangle &\cong \frac{1}{|B|}\int_{B(x,r)} u'\Delta u' \\ &= -\frac{1}{|B|}\int_{B(x,r)} |\nabla u'|^2 + \frac{1}{|B|}\int_{\partial B(x,r)} u'.\frac{\partial u'}{\partial n}\end{aligned}$$

By symmetry (quasi-homogeneous turbulence at the subgrid level) the boundary integral is small (hypothesis H4).
The second term $\nabla.\langle p'u' \rangle$ is treated by a similar argument:

$$\nabla.\langle p'u' \rangle \cong \int_{\partial B(x,r)} p'u'.n \cong 0.$$

Finally the equation for k is found to be

$$\partial_t k + U\nabla k - R : \nabla U - \nabla.(\nu_T \nabla k) + \epsilon = 0.$$

If we use the symmetry of R, the fact that $R : \nabla U = R : (\nabla U + \nabla U^T)/2$ and the formulae for R and ν_T then we obtain

$$\partial_t k + U\nabla k - \frac{c_\mu}{2}\frac{k^2}{\epsilon}|\nabla U + \nabla U^T|^2 - \nabla.(c_\mu \frac{k^2}{\epsilon}\nabla k) + \epsilon = 0.$$

3.3 Derivation of the Equation for ϵ

To obtain an equation for ϵ, one may take the curl of the equation for u', multiply it by $\nabla \times u'$ and use an identity of homogeneous turbulence:

$$\epsilon = \nu\langle|\nabla \times u'|^2\rangle.$$

Letting $\omega' = \nabla \times u'$, its equation is derived from the one of u'

$$\partial_t \omega' + (U + u')\nabla\omega' + u'\nabla\omega - (\omega + \omega')\nabla u' - \omega'\nabla U - \nu\Delta\omega' = -\nabla \times \nabla \cdot R,$$

where $\omega = \nabla \times U$. So after multiplication by $2\nu\omega'$ and averaging, one obtains:

$$0 = 2\nu\langle\omega'.(\partial_t\omega' + (U + u')\nabla\omega' + u'\nabla\omega - (\omega + \omega')\nabla u' - \omega'\nabla U - \nu\Delta\omega')\rangle$$

$$= \partial_t\epsilon + \langle(U + u')\nabla(\nu\omega'^2)\rangle - 2\nu\langle\omega'\nabla \times (u' \times \omega)\rangle$$

$$-2\nu\langle\omega' \otimes \omega'\rangle : \nabla U - 2\nu\langle(\omega' \otimes \omega') : \nabla u'\rangle - 2\nu^2\langle\omega'\Delta\omega'\rangle$$

because

$$\nabla \times (u' \times \omega) = \omega\nabla u' - u'\nabla\omega.$$

The last term is approximated by $2\nu^2\langle|\nabla\omega'|^2\rangle$ thanks to ergodicity and symmetry. The term $\langle u'\nabla\nu\omega'^2\rangle$ is modeled by a diffusion term $-\nabla.(\mu_\epsilon\nabla\epsilon)$ just as in the k equation; for the term $\langle\omega'\nabla \times (u' \times \omega)\rangle$ we have (use ergodicity):

$$\langle\omega'\nabla \times (u' \times \omega)\rangle = \langle(u' \times \omega).\nabla \times \omega'\rangle = -\langle u' \times \omega\Delta u'\rangle$$

$$= -\langle u_2'\Delta u_1'\rangle\omega_3 + \langle u_3'\Delta u_1'\rangle\omega_2 - \langle u_3'\Delta u_2'\rangle\omega_1$$

$$+ \langle u_1'\Delta u_2'\rangle\omega_3 - \langle u_1'\Delta u_3'\rangle\omega_2 + \langle u_2'\Delta u_3'\rangle\omega_1.$$

Thus $\langle\omega'\nabla\times(u'\times\omega)\rangle$ is small, because each term cancels another one approximately; for instance by Green's formula:

$$\langle u_2'\Delta u_1'\rangle = \langle u_1'\Delta u_2'\rangle + \frac{1}{|B|}\int_{B(x,r)}[u_2\frac{\partial u_1}{dn} - u_1'\frac{\partial u_2'}{\partial n}].$$

And these boundary integrals ought to be small by symmetry. Thus far we have

$$\partial_t\epsilon + U\nabla\epsilon - 2\nu\langle\omega'\otimes\omega'\rangle : \nabla U - \nabla.(\mu_\epsilon\nabla\epsilon)$$
$$+ 2\nu\langle(\omega'\otimes\omega') : \nabla u'\rangle + 2\nu^2\langle|\nabla\omega'|^2\rangle = 0.$$

By frame invariance, if $\langle\omega'\otimes\omega'\rangle$ depends only on $\nabla U + \nabla U^T$, k and ϵ (hypothesis H5), then it can only be decomposed on I and $\nabla U + \nabla U^T$ (in two dimensions) and by a dimensionality argument the second term must be proportional to k:

$$\langle\omega'\otimes\omega'\rangle = aI - c_1k(\nabla U + \nabla U^T).$$

The term $\langle(\omega'\otimes\omega') : \nabla u'\rangle$ is neglected (hypothesis H6) because it would be zero if u' was Gaussian (it is a third order moment).
The last term $\nu^2\langle|\nabla\omega'|^2\rangle$ is modeled by a function of k and ϵ. The first polynomial function which is dimensionally correct is ϵ^2/k (hypothesis H7).
Finally one obtains :

$$\partial_t\epsilon + U\nabla\epsilon - \frac{c_1}{2}k|\nabla U + \nabla U^T|^2 - \nabla.(c_\epsilon\frac{k^2}{\epsilon}\nabla\epsilon) + c_2\frac{\epsilon^2}{k} = 0.$$

4. DETERMINATION OF THE CONSTANTS

There are four constants in the model which are adjusted to well understood cases.

4.1 Decay of homogeneous turbulence

When U and ∇U are zero, the model reduces to:

$$\partial_t k + \epsilon = 0, \qquad \partial_t\epsilon + c_2\frac{\epsilon^2}{k} = 0.$$

This is compatible with a polynomial decay in time because $k = k_0(1+\lambda t)^{-n}$, $\epsilon = \epsilon_0(1+\lambda t)^{-m}$ gives

$$-\lambda n k_0 \tau^{-n-1} + \epsilon_0 t^{-m} = 0 \qquad -\lambda m \epsilon_0 \tau^{-m-1} + c_2 \frac{\epsilon_0^2}{k_0} \tau^{n-2m} = 0.$$

with $\tau = 1 + \lambda t$. So

$$m = n + 1 \ = -n + 2m - 1 \quad c_2 = \lambda m \frac{k_0}{\epsilon_0} = \frac{m}{n} = n + \frac{1}{n.}$$

Experiments (Comte-Bellot-Corsin [1966]) give n=1.3 so $c_2 = 2.06$.

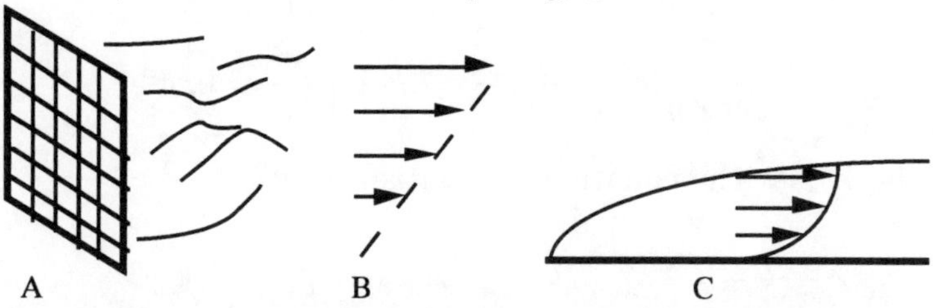

Figure 4.1: The three experiments which determines the constants in the $k - \epsilon$ model. (A) Turbulence behind a grid, (B) shear flow turbulence, (C) turbulence over a flat plate.

4.2 Local Equilibrium Shear layer

If in a frame attached to the mean flow, assumed stationary, we have $\partial_1 U = 0$, $\partial_2 U = C$ constant, then k and ϵ are constant and the equations reduce to

$$c_\mu \frac{k^2}{\epsilon} C^2 = \epsilon, \quad c_1 k C^2 = c_2 \frac{\epsilon^2}{k}.$$

Thus $c_1 = c_\mu c_2$. Measurements of k, ϵ and C give $c_\mu = 0.09$. Therefore $c_1 = c_\mu c_2 = 0.17$.

4.3 Boundary layers

Recall that near solid walls, there may be turbulent boundary layers, for which experimental observation have shown a logarithmic profile for the mean flow [§I.7.3].

More precisely, let $\vec{x}$ be the direction of the flow, y be normal of the wall, and define a velocity scale and a length scale,

$$u^*(x) = \sqrt{\nu \frac{\partial U_1}{\partial y}(x, 0)}, \quad y^* = \frac{\nu}{u^*}.$$

$$\text{Finally let} \quad y^+ = \frac{y}{y^*,} \quad u^+ = \frac{U}{u^*.}$$

Then experiments show that when $20 \leq y^+ \leq 100$ (the so called log layer):

$$u^+ = \frac{1}{\chi}\ln y^+ + 5.5, \qquad \chi = 0.41.$$

In this region the mean flow is stationary and $E = (\partial_y U_1)^2$. So the $k-\epsilon$ equations become

$$-\frac{\partial}{\partial y}(c_\mu \frac{k^2}{\epsilon}\frac{\partial k}{\partial y}) - c_\mu \frac{k^2}{\epsilon}E + \epsilon = 0,$$

$$-\frac{\partial}{\partial y}(c_\epsilon \frac{k^2}{\epsilon}\frac{\partial \epsilon}{\partial y}) - c_1 kE + c_2\frac{\epsilon^2}{k} = 0.$$

Let us check that the equations are satisfied by

$$k = \frac{u^{*2}}{\sqrt{c_\mu}}, \qquad \epsilon = \frac{u^{*3}}{\chi y} \text{ and } E = \frac{u^{*2}}{\chi^2 y^2}.$$

Indeed the two equations above become

$$0 - c_\mu \frac{u^{*4}}{c_\mu}\ \frac{\chi y}{u^{*3}}\ \frac{u^{*2}}{\chi^2 y^2} + \frac{u^{*3}}{\chi y} = 0,$$

$$-\frac{c_\epsilon}{c_\mu}\frac{u^{*4}}{y^2} - c_1\frac{u^{*4}}{\sqrt{c_\mu}}\chi^2 y^2 + c_2\sqrt{c_\mu}\frac{u^{*4}}{\chi^2 y^2} = 0.$$

These are not compatible unless

$$c_\epsilon = \frac{\sqrt{c_\mu}}{\chi^2}(c_2 c_\mu - c_1) = 0.08.$$

Later, these values were modified slightly to give better results. Now most numerical simulations are made with the values given at the beginning of the chapter.

5. BOUNDARY CONDITIONS

Natural boundary conditions could be

$$k, \epsilon \text{ given at t} = 0\ ;\ k|_\Gamma = 0, \epsilon|_\Gamma = \epsilon_\Gamma.$$

However ϵ_Γ is not known at solid boundaries and even if it was it would not be enough because the $k-\epsilon$ model itself is not valid near the solid boundaries where the local Reynolds number (measured with y^+) is not large.

5.1 Wall laws

Originally the $k-\epsilon$ model has been used with wall laws for the parts of Γ corresponding to solid boundaries (see Viollet [1981] for further details). The idea is to remove the viscous sublayer and the log layer of the boundary layer from the computational domain and to use the log law as boundary conditions:

$$k|_{y=\delta} = u^{*2} c_\mu^{-\frac{1}{2}}, \quad \epsilon|_{y=\delta} = \frac{u^{*3}}{\chi\delta}.$$

This works only if $20 \leq \delta/y^* \leq 100$ and when the boundary layer does not separate. Furthermore the computational domain for k and ϵ is not the same as the one for U, unless a wall law is used for it also (see chapter 1). Note that the computational domain is likely to change with time in order to preserve the inequalities $20 \leq \delta/y^* \leq 100$. However the change may be very small in which case it can be neglected. But still the numerical results depend slightly on δ and the above inequalities must be checked a posteriori; if one of them is violated then δ must be changed accordingly.
We have seen in Chapter 3 that wall laws for u are of the type:

$$U.n = 0, \quad \frac{\partial U.s}{\partial n} = g(U.s) \text{ at } y = \delta.$$

Here we know g analytically, almost, because the log law is

$$U.s|\nu\frac{\partial U.s}{\partial n}|^{-\frac{1}{2}} = \frac{1}{\chi}\log(\frac{\delta}{\sqrt{\nu}}|\frac{\partial U.s}{\partial n}|^{\frac{1}{2}}) + \beta.$$

Remark

These boundary conditions are not easy to implement in a computer program (see §III.7.5). They form a highly nonlinear coupling between k, ϵ, U where the Reynolds number appears explicitly (notice the presence of ν above). It is one of the major numerical difficulty of the $k-\epsilon$ model.

5.2 Low Reynolds number $k-\epsilon$ model

Wall-laws avoid the solution of the Navier-Stokes and the turbulence model equations in the near-wall regions. But such a technique fails when separation occurs, at least it is not clear how to take into account recirculating flows.
One possibility is to modify the coefficients of the original $k-\varepsilon$ model to take into account the near-wall damping effects. For this reason,

these models are called "Low-Reynolds". More precisely, in the $k-\varepsilon$ model, the constants c_μ, c_1 and c_2 are multiplied respectively by f_μ, f_1 and f_2 which are positive functions such that $0 < f_\mu \leq 1$, $1 \leq f_1$ and $0 < f_2 \leq 1$ and which depend on two local Reynolds numbers (Hanjalic et al[1976]),

$$R_t = k^2/\nu\varepsilon \text{ and } R_y = \sqrt{k}\frac{y}{\nu}.$$

For instance, in the Lam-Bremhorst [1981] low-Reynolds $k-\varepsilon$ model we have

$$f_\mu = [1 - \exp(-0.0165R_y)]^2(1 + \frac{20.5}{R_t}),$$

$$f_1 = 1 + (\frac{0.05}{f_\mu})^3, \quad f_2 = 1 - exp(-R_t^2),$$

and the wall boundary conditions for u, k and ε are

$$u = 0, \quad k = 0 \quad \text{and} \quad \frac{\partial\varepsilon}{\partial y} = 0.$$

Other model have been proposed (Launder [1992]), but from a finite element implementation point of view, one of the difficulty with these models is that they often require a complex boundary condition for ε on the wall involving somehow the knowledge of the second derivative of k in the direction normal to the wall $(\partial^2 k/\partial^2 y)$. Secondly since ε has strong normal derivatives near the walls these models require a small mesh size near the boundary.

5.3 Coupling with a One-Equation model

Patel et al have concluded [1989] that even for simple configurations, the Low-Reynolds versions of the $k-\varepsilon$ model are not entirely satisfactory. Furthermore, a computation using such models requires considerably more grid points. This is because the turbulent quantities (especially ε) have very large gradients in the sublayer. Therefore, the use of these models in more complex situations may introduce uncertainties and the solution may be mesh dependent. Moreover, often we are not really interested by the exact description of the flow field or the turbulent quantities in the sublayer.

Another solution to avoid near-wall difficulties is to use a two-layer technique (Mohammadi[1992],[1993]). This technique is more complicated than a simple wall law but more practical than a Low-Reynolds two-equations model. Particulary, the model is less mesh dependent and numerically more stable.

The computational region is divided into Ω_H and Ω_L, the high and low Reynolds number regions. In Ω_H the standard $k-\varepsilon$ model is used but in Ω_L the following equation is used for k

$$\frac{\partial k}{\partial t}+u\nabla k-\nabla.((\nu+\nu_T)\nabla k)-\frac{\nu_T}{2}|\nabla u+\nabla u^T|^2+\varepsilon=0$$

and ε is computed by the following algebraic expression

$$\varepsilon=\frac{k^{3/2}}{l_\epsilon}.$$

In this region, The eddy viscosity is given by

$$\nu_T=c_\mu\sqrt{k}l_\mu.$$

In the previous expresions, l_μ and l_ϵ are two length scales which contain the damping effects in the near wall regions

$$l_\mu=\chi {c_\mu}^{-3/4}y(1-\exp(\frac{-y^+}{70}))$$

$$l_\epsilon=\chi {c_\mu}^{-3/4}y(1-\exp(-\frac{y^+}{2\chi {c_\mu}^{-3/4}})).$$

In the previous expression y^+ is defined by $y^+=\frac{\sqrt{k}\rho y}{\mu}$; it depends on the distance y from the wall and on k rather than on u_τ.
Matching of u, k and ε at $\Omega_H\cap\Omega_L$ and $k=0, u=0$ at solid walls closes the model.
At solid walls, the algebraic expression given for ε degenerates, but the value of ε is not needed exactly at the wall.

6. SUMMARY

The $k-\epsilon$ model starts from a Reynolds hypothesis

$$u=U+u',\qquad \langle u'\otimes u'\rangle=-R(k,\epsilon,\nabla U+\nabla U^T)$$

$$\partial_t U+U\nabla U+\nabla P-\nabla.R-\nu\Delta U=0,\quad \nabla.U=0$$

and proposes

$$\nabla.R=c_\mu\nabla.[\frac{k^2}{\epsilon}(\nabla U+\nabla U^T)]$$

where

$$k = \frac{1}{2}\langle|u'|^2\rangle, \qquad \epsilon = \frac{\nu}{2}\langle|\nabla u' + \nabla u'^T|^2\rangle$$

are modeled by

$$\partial_t k + U\nabla k - \frac{c_\mu}{2}\frac{k^2}{\epsilon}|\nabla U + \nabla U^T|^2 - \nabla.(c_\mu \frac{k^2}{\epsilon}\nabla k) + \epsilon = 0,$$

$$\partial_t \epsilon + U\nabla\epsilon - \frac{c_1}{2}k|\nabla U + \nabla U^T|^2 - \nabla.(c_\epsilon \frac{k^2}{\epsilon}\nabla\epsilon) + c_2\frac{\epsilon^2}{k} = 0,$$

with $c_\mu = 0.09$, $c_1 = 0.126$, $c_2 = 1.92$, $c_\epsilon = 0.07$.

The model is derived heuristically from the Navier-Stokes equations with the following hypotheses:
- frame invariance and 2D mean flow, ν_T a polynomial function of k, ϵ.
- u'^2 and $|\nabla \times u'|^2$ are passive scalars when convected by $U + u'$.
- Ergodicity allows statistical averages to be space averages.
- Local homogeneity of the turbulence.
- Reynolds hypothesis for $\langle\omega \otimes \omega\rangle$.
- A closure hypothesis: $\langle|\nabla \times \omega'|^2\rangle = c_2\, \epsilon^2/k$.

The *constants* $c_\mu, c_\epsilon, c_1, c_2$ are chosen so that the model reproduces
- The decay in time of homogeneous turbulence
- The measurements in shear layers in local equilibrium
- The log wall law in boundary layers.

The model is *not valid near solid walls* so it is applied in a computational domain which is within the physical domain at a distance $\delta \in [20, 100]u^*/\nu$.

A possible set of boundary conditions is then

$$k, \epsilon \text{ given at t} = 0 \ ; \quad k|_\Gamma = k_\Gamma, \epsilon|_\Gamma = \epsilon_\Gamma,$$

at artificial inflow and outflow boundaries and

$$k|_{\Gamma+\delta} = |\nu\frac{\partial U.s}{\partial n}|c_\mu^{-\frac{1}{2}}, \quad \epsilon|_{\Gamma+\delta} = \frac{1}{\chi\delta}|\nu\frac{\partial U.s}{\partial n}|^{\frac{3}{2}},$$

at solid walls of normal n and tangent s, and where δ is a function of the tangential coordinates of Γ such that at each point of $\Gamma + \delta$,

$$20\sqrt{\nu|\frac{\partial U.s}{\partial n}|^{-1}} \le \delta \le 100\sqrt{\nu|\frac{\partial U.s}{\partial n}|^{-1}},$$

where $\chi = 0.41, n, s$ are the normal and tangent to the wall.

Alternatives to these boundary conditions have been given in §5.2,5.3. Matching with a one equation model near the wall is actually a better model because it can also handle recirculating flows.

CHAPTER 5

MATHEMATICAL ANALYSIS AND APPROXIMATION

1. INTRODUCTION

It is a tradition in numerical analysis to make a theoretical study of the partial differential system before writing a computer program to solve it. There are many obvious reasons for this such as
- Is the problem well posed (existence and uniqueness)?
- is the solution stable.?

It would indeed be difficult to debug a computer program which would attempt to solve a problem which does not have a solution or has a solution which blows up exponentially.
The $k-\epsilon$ model coupled with the Navier-Stokes equations forms a very complicated systems, not to speak of the boundary conditions; so it may be an impossible task to prove existence, uniqueness and stability. However the model turns out to be rather unstable numerically in certain configuration so it is of practical importance to understand why. There was even some controversy about the existence of a physical solution to the system in configurations with corners.
Now because the viscosity is $\nu + c_\mu k^2/\epsilon$, the model makes no sense if ϵ goes negative. Furthermore the physic requires that k be positive also. So we will analyse this problem first. Then we shall show a partial result on existence and regularity and finally draw some conclusion on the stability of the model and show how it is closely connected to the positivity of k and ϵ.

2. POSITIVITY OF ϵ AND k

For physical and mathematical reasons it is essential that the system of Partial Differential Equations for u, p, k, ϵ yields positive values for k and ϵ.

2.1 Alternative forms
Recall the notations

$$D_t = \frac{\partial}{\partial t} + u\nabla, \qquad E = \frac{1}{2}|\nabla u + \nabla u^T|^2,$$

and the equations for k, ϵ :

$$D_t k - c_\mu \frac{k^2}{\epsilon} E - \nabla.(c_\mu \frac{k^2}{\epsilon}\nabla k) + \epsilon = 0,$$

$$D_t \epsilon - c_1 k E - \nabla.(c_\epsilon \frac{k^2}{\epsilon}\nabla \epsilon) + c_2 \frac{\epsilon^2}{k} = 0,$$

with $c_\mu = 0.09$, $c_\epsilon = 0.07$, $c_1 = 0.126$ and $c_2 = 1.92$.
Now let $\theta = k/\epsilon$. Then

$$D_t\theta = \frac{1}{\epsilon}D_t k - \frac{k}{\epsilon^2}D_t\epsilon = \theta^2 E(c_\mu - c_1) - 1 + c_2$$
$$+ \frac{c_\mu}{\epsilon}\nabla.(\frac{k^2}{\epsilon}\nabla k) - c_\epsilon \frac{k}{\epsilon^2}\nabla.(\frac{k^2}{\epsilon}\nabla\epsilon).$$

Thus

$$D_t\theta = \theta^2 E(c_\mu - c_1) - 1 + c_2 + c_\mu \frac{\theta}{k}\nabla.(k\theta\nabla k) - c_\epsilon \frac{\theta^2}{k}\nabla.(k\theta\nabla\frac{k}{\theta})$$

which can be rewritten as

$$\begin{aligned} &D_t\theta - \theta^2 E(c_\mu - c_1) + 1 - c_2 \\ &\quad = (c_\mu - c_\epsilon)\theta^2\Delta k + c_\epsilon k\theta\Delta\theta + \\ &4\text{sign}(k)(c_\mu - c_\epsilon)\theta^2|\nabla\sqrt{|k|}|^2 + (c_\mu + 2c_\epsilon)\theta\nabla k.\nabla\theta - c_\epsilon k|\nabla\theta|^2. \end{aligned}$$

2.2 Positivity and exponential growth without viscosity
If there were no viscous terms in the equations for k and ϵ then the θ equation would be an autonomous Ordinary Differential Equation on the stream lines:

$$D_t\theta - \theta^2 E(c_\mu - c_1) + 1 - c_2 = 0.$$

It has always a positive bounded solution when the initial and boundary data are positive because $c_\mu < c_1$ and $c_2 > 1$. For example, when E is constant and $\theta(0) = \theta^0$

$$\theta = \frac{\theta^+ - K\theta^- e^{-st}}{1 - Ke^{-st}},$$

where

$$\theta^\pm = \pm\sqrt{\frac{c_2 - 1}{E(c_1 - c_\mu)}}, \quad s = 2\sqrt{(c_2 - 1)(c_1 - c_\mu)E},$$

and

$$K = (\theta^0 - \theta^+)/(\theta^0 - \theta^-).$$

Similarly, in the absence of viscous terms the equation for k reduces to

$$D_t \log k = c_\mu \theta E - \frac{1}{\theta}$$

which has always a positive solution for positive data but it grows exponentially when $c_\mu\theta^2 E > 1$.

2.3 Positivity in the case of Dirichlet boundary data

Let us analyse the full system for θ, k by using the maximum principle [1]. Recall that

$$\begin{aligned} &D_t\theta - \theta^2 E(c_\mu - c_1) + 1 - c_2 \\ &\quad = (c_\mu - c_\epsilon)\theta^2 \Delta k + c_\epsilon k\theta\Delta\theta + \\ &\quad + 4\mathrm{sign(k)}(c_\mu - c_\epsilon)\theta^2|\nabla\sqrt{k}|^2 + (c_\mu + 2c_\epsilon)\theta\nabla k.\nabla\theta - c_\epsilon k|\nabla\theta|^2, \end{aligned}$$

$$D_t k - c_\mu k\theta E - \nabla.(c_\mu k\theta\nabla k) + \frac{k}{\theta} = 0.$$

Assume positive initial data and positive Dirichlet boundary data and suppose that the solution is continuously differentiable.
Let t^* be the first instant for which θ reaches zero and assume that k is positive on $[0, t^*]$. Let x^* be the point where this happens. Because we have assumed θ and k smooth and because x^* cannot be on the boundary (where θ is given), x^* must be a minimum for θ so we have

$$\nabla\theta(x^*, t^*) = 0, \quad \theta(x^*, t^*) = 0, \text{ (and if } \theta \in C^2 : \quad \Delta\theta(x^*, t^*) \geq 0).$$

[1] The basic idea of this proof was given to us by C. Bardos

By writing the θ equation at this point, we obtain :

$$\partial_t \theta = c_2 - 1 > 0$$

This is a contradiction; indeed $t \to \theta(x^*, t)$ has been decreasing up to t^* so $\partial_t \theta was$ negative and suddenly it becomes positive; thus θ is not continuously differentiable (in any case it grows again away from zero). Now let $x^0(t)$ be the minimum of $k(x,t)$ in x. If $x^0(t)$ is on the boundary then the minimum being positive, k is positive at time t. If $x^0(t)$ is not on the boundary then ∇k is zero and $\Delta k \geq 0$ at $\{x^0(t), t\}$. So the k-equation:

$$\partial_t k + U\nabla k - c_\mu k\theta E - c_\mu k\theta \Delta k - c_\mu \nabla(\theta k).\nabla k + \frac{k}{\theta} = 0$$

yields

$$\partial_t k \geq k(c_\mu \theta E - \frac{1}{\theta}) \text{ at } \{x^0(t), t\}.$$

Now let $\kappa(t) = k(x^0(t), t)$. By construction $\partial_t \kappa = \partial_t k$ so the equation above implies

$$\kappa(t) = \min_x k(x,t) \geq \kappa(0) e^{\int_0^t [c_\mu \theta E - \frac{1}{\theta}](x^0(t'), t') dt'}$$

Therefore k is strictly positive.

Remark
If we had $c_\mu = c_\epsilon$ then

$$D_t \theta - \theta^2 E(c_\mu - c_1) + 1 - c_2 = c_\epsilon k\theta \Delta\theta + 3c_\epsilon \theta \nabla\theta.\nabla k - c_\epsilon k |\nabla\theta|^2.$$

So the minimum of θ in x, $q(t)$, satisfies

$$\partial_t q - q^2(c_\mu - c_1)E + 1 - c_2 \geq 0.$$

Now $E \geq 0$. So $\theta \geq q \geq \min\{q(0) + (c_2 - 1)t, q_\Gamma\}$. Thus in this case we have a strictly positive lower bound for θ and k and hence for $k^2/\epsilon = k\theta$. Therefore the system $k - \epsilon$ is *dissipative*.

3. EXISTENCE OF SOLUTION

The previous analysis has one major defect: it assumes that the solution exists and is smooth. It seems hard to prove that it is so.

However Lewandowski et al[1991][1993] have established a partial result for a simpler model. Their analysis is based on a new variable $\varphi = \epsilon^2/k^3$ which satisfies the following

$$
\begin{aligned}
&D_t\varphi + (3c_\mu - 2c_1)E\varphi\theta + (2c_2 - 3)\frac{\varphi}{\theta} \\
&= 2\frac{\epsilon}{k^3}\nabla.(c_\epsilon\frac{k^2}{\epsilon}\nabla\epsilon) - 3\frac{\epsilon^2}{k^4}\nabla.(c_\mu\frac{k^2}{\epsilon}\nabla k) \\
&= -3c_\mu\varphi^2\theta^2\nabla.(\frac{1}{\varphi\theta}\nabla\frac{1}{\varphi\theta^2}) + 2c_\epsilon\varphi^2\theta^3\nabla.(\frac{1}{\varphi\theta}\nabla\frac{1}{\varphi\theta^3}) \\
&= (3c_\mu - 2c_\epsilon)\frac{\Delta\varphi}{\varphi\theta} + 6(c_\mu - c_\epsilon)\frac{\Delta\theta}{\theta^2} + (21c_\mu - 20c_\epsilon)\frac{\nabla\theta.\nabla\varphi}{\theta^2\varphi} \\
&- (9c_\mu - 6c_\epsilon)\frac{|\nabla\varphi|^2}{\varphi^2\theta} - (24c_\mu - 30c_\epsilon)\frac{|\nabla\theta|^2}{\theta^3}.
\end{aligned}
$$

The advantage of this equation over the k-equation is that in the absence of viscous terms (the right hand side), it is explicit in $\log\varphi$:

$$D_t \log\varphi = -(3c_\mu - 2c_1)E\theta - (2c_2 - 3)\frac{1}{\theta}.$$

Hence φ is always decreasing because $3c_\mu - 2c_1 = 0.0188$, and $2c_2 - 3 = 0.84$.

3.1 A model system

Consider the following problem:

$$
\begin{aligned}
&D_t\theta - c_\theta\nabla.(\frac{\nabla\theta}{\varphi\theta}) - E(c_\mu - c_1)\theta^2 + 1 - c_2 = 0 \\
&D_t\varphi - c_\varphi\nabla.(\frac{\nabla\varphi}{\varphi\theta}) + (3c_\mu - 2c_1)E\varphi\theta + (2c_2 - 3)\frac{\varphi}{\theta} = 0,
\end{aligned}
$$

where a, b, c, e are smooth C^∞ functions of x, t strictly positive. and where c_φ and c_θ are strictly positive constants. Consider the following boundary conditions:

$$
\begin{aligned}
\theta(0) = \theta^0 \quad & \varphi(0) = \varphi^0 \\
\theta|_\Gamma = \theta_\Gamma \quad & \varphi|_\Gamma = \varphi_\Gamma
\end{aligned}
$$

This model system is the exact $k - \epsilon$ system with a modified diffusion of the same type as the original one because $k^2/\epsilon = 1/\varphi\theta$. A similar idea was introduced independently by Coakley[1983] for his k, ω system where $\omega = \epsilon/k = 1/\theta$. This model system is mathematically attractive and it contains all the difficulties of the original one, however our numerical experiments on this model show that it is too dissipative.

Theorem (Lewandowski)
Assume that $E, u \in L^\infty(\Omega\times]0,T'[)$ *and that* $\theta^0, \varphi^0, \theta_\Gamma, \varphi_\Gamma$ *are strictly positive and bounded. Then there exists* $\{\theta, \varphi\} \in L^2(0,T'; H^1(\Omega))^2$ *solution of the model system* [2] *Furthermore* θ *and* φ *are positive and bounded.*

Proof: see Appendix A5.

4. NUMERICAL METHODS

4.1 Variational formulation

Consider the following model :

$$\partial_t u + u\nabla u + \nabla p - \nabla.[\nu_T(\nabla u + \nabla u^T)] = 0, \quad \nabla.u = 0$$

$$u.n = 0, \quad au.s + \nu\frac{\partial u.s}{\partial n} = b$$

where a, b are given functions of $u, k, \epsilon, \nabla u$... and $\nu_T = \nu + c_\mu k^2/\epsilon$. This mixed boundary condition comes from a wall law, as explained in §III.7.5.

The variational formulation can be written in the space of functions having zero divergence and having zero normal trace:

$$(\partial_t u, v) + (u\nabla u, v) + \frac{1}{2}(\nu_T(\nabla u + \nabla u^T), \nabla v + \nabla v^T) + \int_\Gamma [au.v - bv] = 0$$

$$\forall v \in J_{on}(\Omega); \quad u \in J_{on}(\Omega) = \{v \in H^1(\Omega)^3 : \nabla.v = 0, v.n|_\Gamma = 0\}$$

4.2 Discretization with Galerkin Characteristics

By discretizing the total derivative, we can consider a semi-implicit scheme,

$$\frac{1}{\delta t}(u_h^{m+1} - u_h^m o X_h^m, v_h) + \frac{1}{2}(\nu_{Th}^m(\nabla u_h^{m+1} + \nabla u_h^{m+1^T}), \nabla v_h + \nabla v_h^T)$$

$$+ \int_\Gamma (a^{m+1}u_h^m - b^m)v_h d\gamma = 0, \quad \forall v_h \in J_{onh}$$

2 Uniqueness is conjectured only.

$$u_h^{m+1} \in J_{onh} = \{v_h \in V_h : (\nabla . v_h, q_h) = 0 \quad \forall q_h \in Q_h; \quad v_h . n_h|_\Gamma = 0\}.$$

where V_h and Q_h are as in Chapter 3 and where ν_{Th}^m is ν_T, evaluated at time $n\delta t$.
There is a difficulty with J_{onh} and the choice of the approximated normal n_h, especially if Ω has corners but we can also replace J_{onh} by

$$J'_{onh} = \{v_h \in V_h : \quad (v_h, \nabla q_h) = 0 \quad \forall q_h \in Q_h\}$$

because

$$0 = (u, \nabla q) = -(\nabla . u, q) + \int_\Gamma u.nqd\gamma, \forall q \quad \Rightarrow \quad \nabla . u = 0 \text{ and } u.n|_\Gamma = 0.$$

With J'_{onh} the slip boundary conditions are satisfied in a weak sense only, but the normal n_h does not appear now.
The techniques developed for Navier-Stokes equations can be adapted to this framework, in particular the solution of the linear systems can be carried out with the conjugate gradient algorithm (see Glowinski[1987] or Pironneau[1989] for example) or, preferably, with a bi-conjugate gradient method. However we note that the matrices would have to be reconstructed at each iteration because a, ν_T depend on m.

To solve the equations $k - \epsilon$, we can use the same method ; we add to the previous system: for all $w_h \in W_{oh}$

$$(k_h^{m+1} - k_h^m o X_h^m, w_h) + \delta t c_\mu (\frac{k_h^{m^2}}{\epsilon_h^m} \nabla k_h^{m+1}, \nabla w_h)$$

$$+ (\int_{m\delta t}^{(m+1)\delta t} [\epsilon_h^m - c_\mu \frac{k_h^{m^2}}{\epsilon_h^m} E_h^m](X(t))dt, w_h) = 0,$$

$$(\epsilon_h^{m+1} - \epsilon_h^m o X_h^m, w_h) + \delta t c_\epsilon (\frac{k_h^{m^2}}{\epsilon_h^m} \nabla \epsilon_h^{m+1}, \nabla w_h^m) +$$

$$-(\int_{m\delta t}^{(m+1)\delta t} [c_1 k_h^m E_h^m - c_2 \frac{\epsilon_h^{m^2}}{k_h^m}](X(t))dt, w_h) = 0,$$

where $E_h^m = |\nabla u_h^m + \nabla u_h^m|^2/2$.
The integrals from mk to $(m+1)k$ are carried out along the streamlines in order to stabilize the numerical method (Goussebaile-Jacomy [1985]).

So, at each iteration, we must compute the right hand sides by the Characteristic-Galerkin method and
- solve the linear system for u_h^{m+1}, p_h^{m+1},
- solve the linear system for k_h^{m+1} and the one for ϵ_h^{m+1}.

The algorithm is not very stable and converges slowly in some cases. Much improvement is obtained if in a more implicit version both equations are solved simultaneously by a quasi-Newton method such as GMRES:

$$(k_h^{m+1} - k_h^m o X_h^m, w_h) + \delta t c_\mu(\frac{k_h^{m^2}}{\epsilon_h^m} \nabla k_h^{m+1}, \nabla w_h)$$
$$+ (\int_{m\delta t}^{(m+1)\delta t} [\epsilon_h^{m+1} - c_\mu \frac{k_h^{m+1^2}}{\epsilon_h^{m+1}} E_h^m](X(t))dt, w_h) = 0,$$

$$(\epsilon_h^{m+1} - \epsilon_h^m o X_h^m, w_h) + \delta t c_\epsilon(\frac{k_h^{m^2}}{\epsilon_h^m} \nabla \epsilon_h^{m+1}, \nabla w_h^m)$$
$$- (\int_{n\delta t}^{(m+1)\delta t} [c_1 k_h^{m+1} E_h^m - c_2 \frac{\epsilon_h^{m+1^2}}{k_h^{m+1}}](X(t))dt, w_h)$$
$$= 0, \text{ for all } w_h \in W_{oh}.$$

Our experience has shown to us that it is not necessary to couple in a fully implicit fashion the equations for u_h^{m+1}, p_h^{m+1} with those for k_h^{m+1} and ϵ_h^{m+1} . But it is important to couple the equation for k with the equation for ϵ and solve them simultanously.

4.3 A simple semi-implicit scheme

To reduce the cost of the fully implicit scheme consider the following method:

$$(k_h^{m+1} - k_h^m o X_h^m, w_h) + \delta t c_\mu(\frac{k_h^{m^2}}{\epsilon_h^m} \nabla k_h^{m+1}, \nabla w_h) + \delta t(k_h^{m+1} \frac{\epsilon_h^m}{k_h^m}, w_h)$$
$$= \delta t c_\mu(\frac{k_h^{m^2}}{\epsilon_h^m} E_h^m, w_h)$$

$$(\epsilon_h^{m+1} - \epsilon_h^m o X_h^m, w_h) + \delta t c_\epsilon(\frac{k_h^{m^2}}{\epsilon_h^m} \nabla \epsilon_h^{m+1}, \nabla w_h^m) + \delta t c_2(\epsilon_h^{m+1} \frac{\epsilon_h^m}{k_h^m}, w_h)$$
$$= \delta t c_1(k_h^m E_h^m, w_h), \text{ for all } w_h \in W_{oh}.$$

The basic idea is to split the terms of order zero into their positive part and negative part and treat implicitly the positive terms and explicitly the other ones. Then all terms on the left hand side are positive and so are all terms on the right hand side. The maximum principle for PDE in the discrete case insures positive $k_h^{m+1}, \epsilon_h^{m+1}$ when the triangulations have sharp angles only.

4.4 A stable semi-implicit multi-step scheme

The previous to last scheme is numerically expensive because a few Newton loops are necessary every time step. The last scheme yields positive values but it is not very stable. By using what we know about the positivity of the $k-\epsilon$ model, it is possible to build a stable multistep scheme which involves only linear systems at each time step also and is more stable.

At every time step, the Navier-Stokes equations are solved with ν_T and the boundary conditions computed at the previous time step. The equations for $k-\epsilon$ are solved by a multistep algorithm involving one step of convection and one step of diffusion. However in this case the convection step is performed on k, θ or φ, θ rather than on k, ϵ.

The equation for θ is integrated without diffusion:

$$\begin{aligned}&(\theta_h^{m+\frac{1}{2}}, w_h) + (\theta_h^m \theta_h^{m+\frac{1}{2}} E_h^m, w_h)[c_1 - c_\mu]\delta t \\ &= (\theta_h^m o X_h^m, w_h) + (c_2 - 1, w_h)\delta t\end{aligned}$$

with $\theta_h^m = k_h^m / \epsilon_h^m$. The equation for $\varphi = \epsilon^2/k^3$ (see §3) is also integrated without diffusion:

$$(\varphi_h^{m+\frac{1}{2}}, w_h) + \delta t((3c_\mu - 2c_1)E_h^m \varphi_h^{m+\frac{1}{2}} \theta_h^m, w_h)$$

$$+\delta t((2c_2 - 3)\frac{\varphi^{m+\frac{1}{2}}}{\theta_h^{m+\frac{1}{2}}}, w_h) = (\varphi_h^m o X_h^m, w_h), \quad \forall w_h \in W_h.$$

Then $k_h^{m+1/2}, \epsilon_h^{m+1/2}$ are computed from the formulae

$$k_h^{m+\frac{1}{2}} = \frac{1}{\varphi_h^{m+\frac{1}{2}} (\theta_h^{m+\frac{1}{2}})^2}, \qquad \epsilon_h^{m+\frac{1}{2}} = \frac{k_h^{m+\frac{1}{2}}}{\theta_h^{m+\frac{1}{2}}}.$$

Then the diffusion step is applied to k and ϵ,

$$(k_h^{m+1}, w_h) + \delta t c_\mu (\frac{k_h^{m^2}}{\epsilon_h^m} \nabla k_h^{m+1}, \nabla w_h) = (k_h^{m+\frac{1}{2}}, w_h),$$

$$(\epsilon_h^{m+1}, w_h) + \delta t c_\epsilon (\frac{k_h^{m^2}}{\epsilon_h^m} \nabla \epsilon_h^{m+1}, \nabla w_h) = (\epsilon_h^{m+\frac{1}{2}}, w_h),$$

for all $w_h \in Q_{oh}$; $\epsilon_h^{m+1} - \epsilon_{\Gamma h} \in W_{oh}$, $k_h^{m+1} - k_{\Gamma h} \in W_{oh}$.

Proposition
With Lagrangian Finite Element of degree 1 on a triangulation without obtuse angles and with mass lumping on the first and last integrals in the diffusion step, the above scheme cannot produce negative values for k_h^{m+1} *and* ϵ_h^{m+1} .

Proof : Each step produces positive values only. It is known (Ciarlet [1978]) that the maximum principle holds in the discrete case, with P^1 finite elements and triangles with sharp angles, for coercive operator, like the one in the diffusion step.

4.5 Matching different models at the walls

We have seen in Chapter 4 that the $k - \epsilon$ model is integrated with very complex boundary conditions. Three strategies were given:

- Wall laws
- Low Reynolds number extentions
- One-equation models or algebraic models near solid walls.

Wall laws usually result in Dirichlet conditions for k and ϵ. There are no programming difficulties there. Low Reynolds number extentions are also easy because the boundary conditions are $k = 0, \epsilon = \epsilon_0$. They require however a very fine mesh near the walls in order to resolve the strong gradients.
The third approach seems difficult at first sight but is practically very simple because of the semi-implicit aspect of the algorithms given *above*.

For example consider the case where near the walls the equation for k is kept and the equation for ϵ is replaced by an algebraic formula $\epsilon(x) = \lambda(y)k(x)^{3/2}$ where λ is a given function of the distance between x and the nearest wall.

Now take first the case of algorithm 4.3 described above. At the mth time step, k^m, ϵ^m are known. We assume that the zone where the algebraic expression for ϵ is to be used has been identified. Since ϵ^{m+1} does not appear in the equation for k^{m+1}, the later can be computed without change. For ϵ^{m+1} one identifies the set of vertices of the mesh which belong to the domain where the ϵ equation must be solved and use on all other vertices the value $\epsilon^{m+1} = \lambda(k^{m+1})^{3/2}$. In practice this change of domain can be implemented by penalty, i.e. by redefining the diagonal element of the linear system for ϵ^{m+1} as 10^{20} and putting on the same row in the right hand side $10^{20}\lambda(k^{m+1})^{3/2}$. Finally the zones are redefined from the values of k^{m+1} and ϵ^{m+1} and the next time step is investigated.

In the case of Algorithm 4.4, it is the same thing. Convection is performed without change. Then diffusion for k^{m+1} is done also without change and it is only for the diffusion step for ϵ^{m+1} that the matrix of the linear system must be modified on its diagonal by blocking into a Dirichlet state all elements corresponding to vertices in the zone where ϵ is defined algebraically from k.

5. PERFORMANCES AND NUMERICAL RESULTS

Considering the number of hypotheses that were necessary to derive the model, it performs surprisingly well, even on flows which do not satisfy the asumptions. Rather than list the cases on which it performs properly, we give a few counter examples here and in the next chapter. We also give some numerical results on simple test cases.

The $k-\epsilon$ model is really meant for flows without recirculation. However it performs fairly well also on some recirculating flows. For boundary layers, it does not necessarily improve on the classical mixing length algebraic boundary layer models (Cousteix[1990]). For 3D flows, there are a number of cases where it has been shown not to do well in particular when the stress tensor and the Reynolds tensor are not parallel (hypothesis (H1)) (Speziale[1988], see also chapter 10). For time dependent mean flows it may also not be so good (Begue et al[1990], Rodi[1986]).

In the following we report on some simple and classical test cases. All except the Mixing Layer have been computed with a finite element method based on the Q^2/P^1 quadrilateral element described in Chapter 3; upwinding is implemented via the Characteristic-Galerkin method. The mixing layer was computed with the program for compressible flow described in Chapter 6 (finite-volumes / finite-elements) for the part concerning u, p and by the algorithms of §4.2,4.3 for k and ϵ. The Mach number was set to a small value in order to have an incompressible fluid.

5.1 Grid Turbulence

Turbulence behind a grid corresponds to a uniform flow $u = \{1,0,0\}^T$ which convects a homogeneous turbulence $k = k^0, \epsilon = \epsilon^0$. It is very similar to the decay of homogeneous turbulence as discussed in §IV.4.1 except that t is replaced by x_1.
There is analytical solution:

$$k = k^0(1 + (c_2 - 1)x_1 \frac{\epsilon^0}{k^0})^{\frac{1}{1-c_2}}, \quad \epsilon = \epsilon^0(1 + (c_2 - 1)x_1 \frac{\epsilon^0}{k^0})^{\frac{c_2}{1-c_2}}.$$

It is a good numerical test to validate a computer program for $k - \epsilon$. Figures 5.1, 5.2 give the numerical results with the algorithm of §4.4.

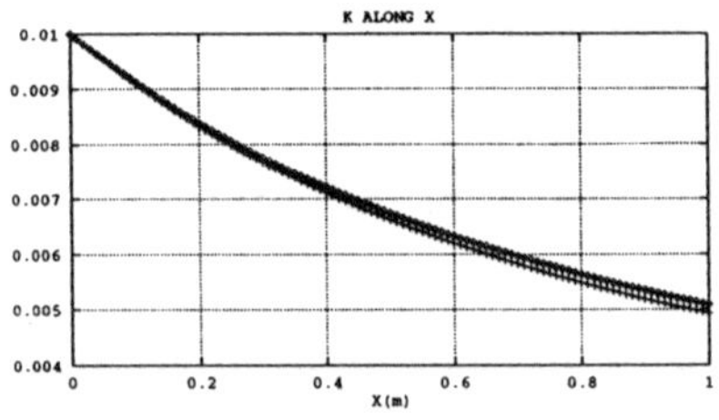

Figure 5.1 :Computed and theoretical k versus x_1 for a turbulence behind a grid

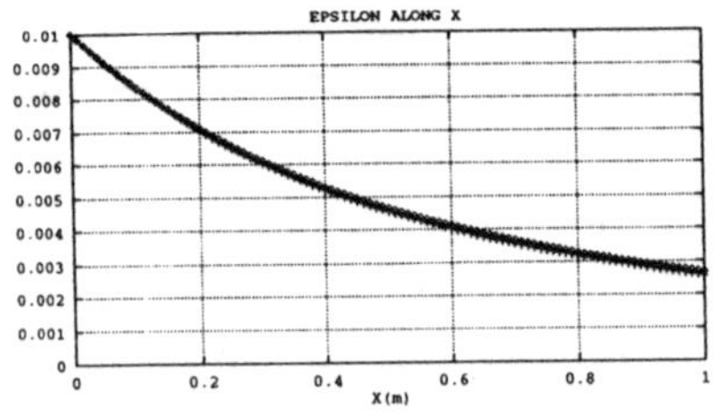

Figure 5.2 :Computed and theoretical epsilon versus x_1 for a turbulence behind a grid

5.2 Mixing Layer

The problem is in two dimensions and the computational domain is a rectangle $]0, L[\times] - l, +l[$. The mesh has 1200 triangles and is refined around the line $x_2 = 0$.
The mean flow is given at time 0:

$$u^0 = \begin{cases} \{1,0\} & \text{when } y \geq 0 \\ \{2,0\} & \text{when } y < 0 \text{ ;} \end{cases}$$

k^0 and ϵ^0 are set to some small constant values . The boundary conditions are $(u^0, k^0\epsilon^0)|_\Gamma$ given except at the exit boundary where a Neumann condition is applied.
The results have been obtained with the algorithm of §4.2(figures 5.3) and with the one of §4.4 (figures 5.4). Figure 5.5 shows that the time step for the algorithm was chosen too big because the solution becomes time dependent after a while; it shows also the superiority of algorithm 4.4.

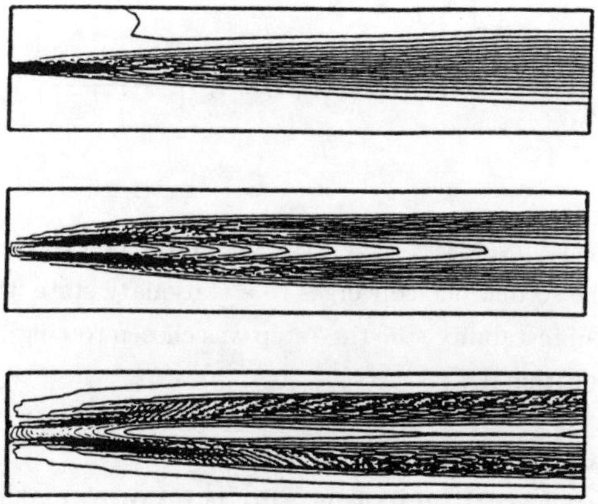

Figure 5.3 : Mixing layer. Top shows k, middle epsilon, bottom turbulent viscosity. The results have been obtained with algorithm of section 4.2 .

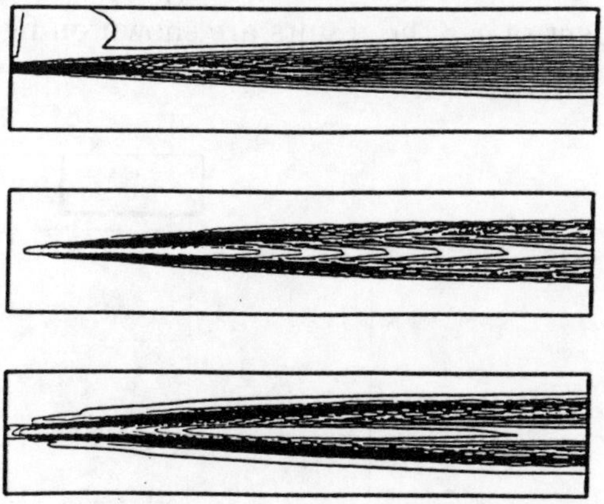

Figure 5.4 : Mixing layer. Top shows k, middle epsilon , bottom turbulent viscosity. The results have been obtained with algorithm of section 4.4.

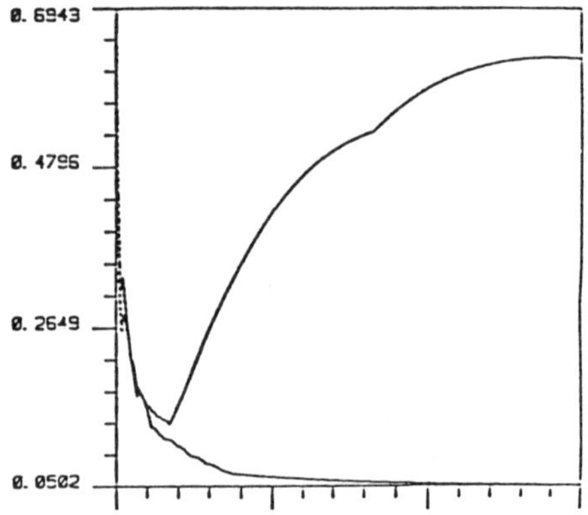

Figure 5.5 : Residual curve $\epsilon^{m+1} - \epsilon^{m}$ versus m for the two computations of figures 5.3 and 5.4. While the second one converges to a stationary state, the first one does not. This is an indication of instability (the time step was chosen too big); it also demonstrates the superiority of algorithm 4.4.

5.3 Flow between two flat plates

The problem is in two dimensions and the computational domain is a rectangle $]0, L[\times]-l, +l[$. The mesh has 200 quadrangles. The mean flow at time zero is 0, $k^0 = 0.5$, $\epsilon^0 = 1$. The boundary conditions at x=0 are set to the experimental valued measured by Comte-Bellot [1980]. At the exit boundary a Neumann condition is applied to u, k, ϵ. On the plates $x_2 = 0$, $x_2 = l$, a wall law (§IV.5.1) is applied with Re=57000.
The results have been obtained with the algorithm of §4.2. The flow is rapidly independent of x_1; the results are shown on figure 5.6.

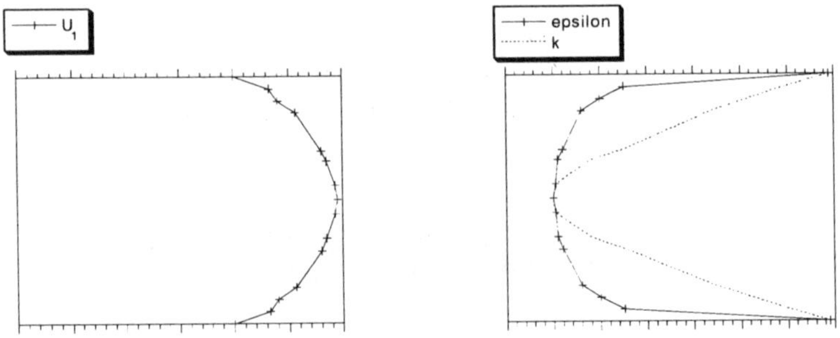

Figure 5.6 (computed by B. Cardot[1989]): Flow between two flat plates. The graph shows the asymptotic values of u_1,k and epsilon for large x_1. The max of epsilon is 35.5,

its minimum is 0.5 and k belongs to [0.05,0.1], u belongs to [0.68,1]. Turbulence is stronger near the walls. In this computation relatively few points were used, so as to test robustness for more complex cases in the future.

5.4 Flow in a cavity

The domain is a square with a longer rectangle tangent to the top . A quadrilateral mesh with 368 elements is used, which is again purposely little (see chapter 9 for a better computation). The fluid flows from left to right and creates an eddy in the cavity. At time zero all variables are set to zero except k and ϵ which are set to a small value. The inflow boundary is the first vertical segment on the right; the boundary conditions are the results of the computation above "flow between two parallel plates", except that the Reynolds number is 10^4. On the outflow boundary (the last vertical segment on the right), Neumann conditions are imposed. At the walls a wall-law is prescribed. Results are shown on figure 5.7.

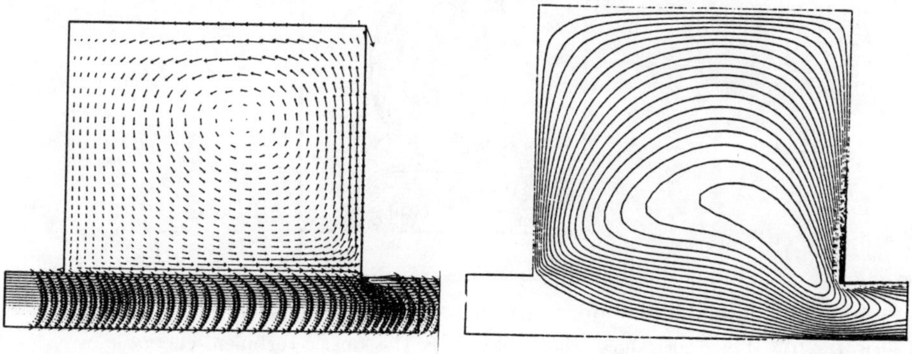

Figure 5.7 : Flow in a cavity at Re=10^4 with a wall law at the solid boundaries, Dirichlet conditions on the inflow boundary (top left) corresponding to turbulent pipe flow and Neumann conditions on the top-right vertical boundary. On the left the figure shows the velocities at the vertices of the mesh, on the right the level lines of the turbulent diffusion

5.5 Transient flow around a cylinder

The domain is R^2 minus a circle (cylinder) and an artificial boundary approximates infinity. It is quadrangulated with 660 quadrilaterals. The flow is uniform at infinity and equal to $\{1,0,0\}^T$. A wall law is imposed on the cylinder (which by the way is unrealistic) with $Re = 10^4$. At infinity $k = \epsilon = 10^{-7}$. Initial conditions are $u = 0, k = 0.5, \epsilon = 1$ so as to have a smooth start in the algorithm. Algorithm of §4.2 did not converge while the one of §4.4 gave a transient solution (the stationary solution is unstable) . Figure 5.8 shows k. Notice that the turbulence model elongates the main eddies; a phenomenon which was not observed experimentally.

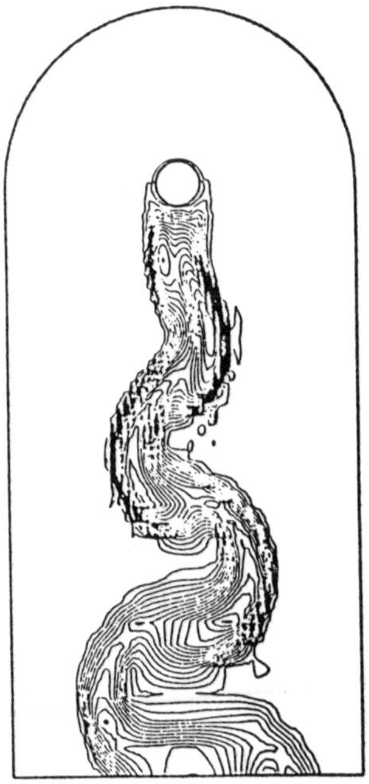

Figure 5.8 : Transient flow around a cylinder with k-epsilon, a wall law on the cylinder with Re=10^4. The figure shows the level lines of the kinetic turbulent energy k.

APPENDIX A5

AN EXISTENCE THEOREM FOR THE REDUCED PHI-THETA MODEL

Consider the following problem:

$$D_t\theta - c_\theta \nabla.(\frac{\nabla\theta}{\varphi\theta}) + e'E\theta^2 - c = 0$$

$$D_t\varphi - c_\varphi \nabla.(\frac{\nabla\varphi}{\varphi\theta}) + a'E\varphi\theta + b\frac{\varphi}{\theta} = 0$$

where a', b, c, e' are strictly positive constants. Consider the following boundary conditions:

$$\theta(0) = \theta^0 \quad \varphi(0) = \varphi^0$$

$$\theta|_\Gamma = \theta_\Gamma \quad \varphi|_\Gamma = \varphi_\Gamma.$$

Theorem (Lewandowski[1992])
Assume that $u \in L^\infty(\Omega\times]0,T'[)$ *and that* $E, \theta^0, \varphi^0, \theta_\Gamma, \varphi_\Gamma$ *are strictly positive and bounded. Then there exists* $\{\theta, \varphi\} \in L^2(H^1(\Omega))^2$ *solution of the model system. Furthermore* θ *and* φ *are positive and bounded.*

Proof (Sketch).
Denote $a = Ea', e = Ee'$ and

$$\theta_M = \max\{1, \sup_\Omega \theta^0\} \quad \theta_m = \inf_{\Omega\times]0,T'[}\{\theta^0(x), \sqrt{\frac{c}{E(x,t)e'}}\}$$

$$\varphi_M = \sup_\Omega \varphi^0 \quad \varphi_m = \inf_\Omega \theta^0.$$

Let

$$\theta' = \begin{cases} \theta_m \text{ if } \theta \le \theta_m \\ \theta_M \text{ if } \theta \ge \theta_M \\ \theta \text{ otherwise} \end{cases}$$

and let φ' be defined similarly. Consider now the regularized system

$$D_t\theta - c_\theta \nabla.(\frac{\nabla\theta}{\varphi'\theta'}) = -\theta'^2 e + c$$

$$D_t\varphi - c_\varphi \nabla.(\frac{\nabla\varphi}{\varphi'\theta'}) = -\varphi'(a\theta' + \frac{b}{\theta'})$$

with the same initial and boundary conditions.
The proof proceeds in two steps:
First one shows that any solution of the regularized problem satisfies $\theta = \theta'$, $\varphi = \varphi'$. This of course implies that any solution of the regularized problem is a solution of the original problem.
Then one shows that the regularized problem has one and only one solution.

Lower bound for θ
As usual $f^\pm$ denotes the positive and negative part of f ($f = f^+ - f^-$). Multiply the $\theta-$ equation by $-(\theta-\theta_m)^-$ and integrate. Since for any spatial derivative,

$$-\partial\theta(\theta-\theta_m)^- = \frac{1}{2}\partial|(\theta-\theta_m)^-|^2,$$

one has

$$D_t\int_\Omega |(\theta-\theta_m)^-|^2 + 2\int_\Omega \frac{c_\theta}{\varphi'\theta'}|\nabla(\theta-\theta_m)^-|^2 = -2\int_\Omega (c-\theta_m^2 e)(\theta-\theta_m)^-.$$

The second integral is the result of Green's formula and there are no boundary terms because $(\theta-\theta_m)^-|_\Gamma = 0$.
Now by definition, $c - \theta_m^2 e \geq 0$, so the right hand side is negative and one is left with

$$D_t\int_\Omega |(\theta-\theta_m)^-|^2 \leq 0.$$

Thus this integral is non increasing. But it is zero at $t = 0$ and always positive, so $\theta \geq \theta_m$.

Upper bound for θ
Exactly the same argument is repeated with $(\theta-\theta_M)^+$. One obtains

$$D_t\int_\Omega |(\theta-\theta_M)^+|^2 + 2\int_\Omega \frac{c_\theta}{\varphi'\theta'}|\nabla(\theta-\theta_M)^+|^2 = 2\int_\Omega (c-\theta_M^2 e)(\theta-\theta_M)^+$$

This shows that $\theta \leq \theta_M$ because $c - \theta_M^2 e$ is negative.

Upper bound for φ

Next with the equation for φ and $(\varphi-\varphi_M)^+$

$$D_t\int_\Omega |(\varphi-\varphi_M)^+|^2 + 2\int_\Omega \frac{c_\varphi}{\varphi'\theta'}|\nabla(\varphi-\varphi_M)^+|^2$$
$$= -2\int_\Omega \varphi_M(a\theta'+\frac{b}{\theta'})(\varphi-\varphi_M)^+$$

The right hand side being negative we conclude that $\varphi\le\varphi_M$ at all times.

Lower bound for φ

Let φ_0,λ be two constants and consider $\psi=(\varphi-\varphi_0 e^{-\lambda t})$

$$-\int_\Omega \psi^-[D_t\varphi - c_\varphi\nabla(\frac{\nabla\varphi}{\varphi'\theta'})] = \int_\Omega[\frac{1}{2}D_t\psi^{-^2} + \lambda\varphi_0 e^{-\lambda t}\psi^- + \frac{c_\varphi}{\varphi'\theta'}|\nabla\psi^-|^2]$$
$$= \int_\Omega \psi^-(a\theta'+\frac{b}{\theta'})\varphi'.$$

So

$$D_t\int_\Omega \psi^{-^2} \le 2\int_\Omega \psi^-(a\theta'+\frac{b}{\theta'}-\lambda\varphi_0 e^{-\lambda t})\varphi'$$

The right hand side will be negative if $\varphi'(a\theta'+b/\theta')\le -\lambda\varphi_0 e^{-\lambda t}$. But if $\varphi_M\le\varphi_0 e^{-\lambda t}$ then $\varphi\le\varphi_0 e^{-\lambda t}$ and the inequality will be satisfied if $\lambda\ge\max_{z\in(\theta_m,\theta_M)}(az+b/z)$.

Now existence and uniqueness is shown by a fixed point theorem. For clarity let $\theta_\Gamma=\varphi_\Gamma=0$.

Recall that $f\to G(f)=\int_0^t f(x,\tau)d\tau$ is a linear map from $H=L^2(H^1)$ into H and that $\nabla G(f)=G(\nabla f)$ with $||G(f)||\le||f||_H$.

Let $V=H\times H$; define $F:V\to V'$ by

$$F(\theta,\varphi)=\begin{pmatrix}\theta^0+\int_0^t(c-\theta'^2 e)-u\nabla G(\theta)+\int_0^t\nabla.(\frac{c_\theta}{\theta'\varphi'}\nabla\theta')\\ \varphi^0+\int_0^t(a\theta'+\frac{b}{\theta'})\varphi'-u\nabla G(\varphi)+\int_0^t\nabla.(\frac{c_\varphi}{\theta'\varphi'}\nabla\varphi)\end{pmatrix}.$$

Clearly

$$
\begin{aligned}
&||F(\theta,\varphi)|| \\
&\leq [(||\theta^0|| + T' f_1(\theta_m,\theta_M,\varphi_m,\varphi_M) \\
&+ ||\theta||_H[||u||_\infty + \frac{c_\theta}{\theta_m\varphi_m}])^2 + (||\varphi^0|| + T' f_1(\theta_m,\theta_M,\varphi_m,\varphi_M) \\
&+ ||\varphi||_H[||u||_\infty + \frac{c_\varphi}{\theta_m\varphi_m}])^2]^{\frac{1}{2}}
\end{aligned}
$$

So if $||\theta,\varphi|| \leq R$ then $||F(\theta,\varphi)|| \leq \alpha + \beta R$ with $\beta = ||u||_\infty + (c_\theta + c_\varphi)/\theta_m\varphi_m$.
Thus F is continuous and contracting when R is large enough, i.e. when $\beta < 1$. The Leray-Schauder theorem applies. To ensure this last constraint a change of scale $x \to rx$ is performed. It changes $u \to ru$, $c_\theta, c_\varphi \to r^2 c_\theta, r^2 c_\varphi$. So β is made less than one by this operation.

Remark
Along the way the following has been shown:

$$\min\{\inf\sqrt{\frac{c}{e}}, \inf\theta^0\} \leq \theta \leq \max\{\sup\sqrt{\frac{c}{e}}, \sup\theta^0\}$$

$$\inf\theta^0 e^{-\lambda T'} \leq \varphi \leq \sup\varphi^0$$

where $\lambda = \inf_x \max_{z\in(\theta_m,\theta_M)}(az + b/z)$.

CHAPTER 6

OTHER MODELS BEYOND K-EPSILON

1. INTRODUCTION

In this chapter, we give a brief description of some turbulence models beyond $k-\varepsilon$. We do not describe here the modeling process but rather the numerical difficulties which appears with these models.
As was said earlier the $k-\epsilon$ model performs rather well but due to Reynolds' hypothesis (the Reynolds stress tensor parallel to the deformation tensor), such two-equations models have some deficiencies. The most important are:

1. their inability to account for rotational strain or shear,
2. their inaccurate prediction of normal Reynolds stress anisotropies,
3. their inability to account for the amplification or the relaxation of the components of the Reynolds stress tensor.

We will give a few examples about each of these cases.
Concerning the first point, it is well known that the $k-\varepsilon$ model fails to distinguish between the case of plane strain, plane shear[1]
and rotating plane strain or shear (see Speziale[1988]).
Indeed, the k and ε predicted by this model are independent of the state of rotation of the fluid. This fact is in clear contradiction with results from direct numerical simulation of the Navier-Stokes equations or experimentations.

1

- Plain strain: there exists $\partial_i u_i >> \partial_j u_i$ for all $j \neq i$.
- Plain shear: there exists $\partial_i u_i << \partial_j u_i$ for all $j \neq i$
- Rotational plane strain: = plain strain + rotation

In the same way, the $k-\varepsilon$ model over estimates the turbulence levels in situations where the flow is perpendicular to a wall. This is due to the fact that the model does not distinguish between shear and strain having the same distribution of $|\nabla u+\nabla u^T|^2$. For example, in the simulations of flows around cylinders, the $k-\varepsilon$ model predicts very high levels of turbulence before the cylinder even in the subcritical range $Re < 2.10^5$ where the boundary layer remains laminar before the separation points Achenbach[1968].

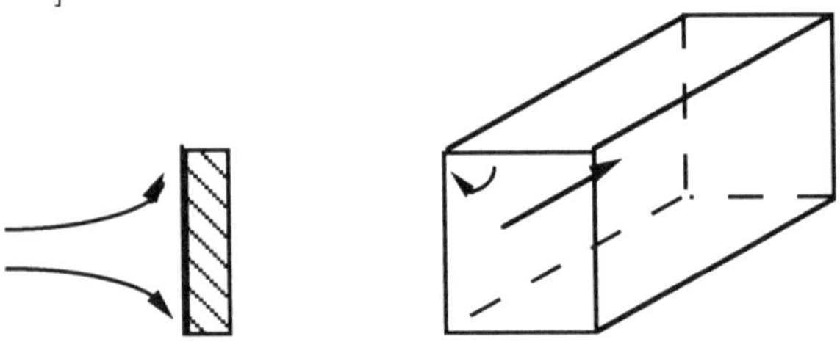

Figure 6.1:Two cases where $k-\epsilon$ fails. Left: flow against a perpandicular wall. Right: flow in a channel with square cross section where the secondary eddies near the corners are not predicted by the model.

Another example where the $k-\varepsilon$ fails, consists of a unidirectional mean flows in ducts with polygonal cross-section. Indeed, the model does not produce secondary eddies near corners, in contradiction with experimental results (Speziale[1987]). This deficiency comes also from Reynolds' hypothesis which in the $k-\varepsilon$ model gives in this case:

$$R_{11}=R_{22}=R_{33}.$$

Finally, for the third point, classical two-equations models often give erroneous predictions for non-equilibrium flows (e.g. flows with amplification or a relaxation of turbulence). For instance, for an initially anisotropic turbulence behaving in a region where the velocity gradients are set to zero, the $k-\varepsilon$ model predicts an instantaneous return to isotropy because the Reynolds stress is reduced to

$$R_{ij}=-\frac{2}{3}k\delta_{ij}.$$

However, experiments indicate that this return to isotropy is gradual. In summary, all of these deficiencies seem to come from the assumption that Reynolds' stress tensor is parallel to the deformation tensor and that the local state of turbulence is only characterized by an isotropic velocity, $k^{1/2}$, and length scale, $k^{3/2}/\varepsilon$.

To overcome this difficulty, several other modelization for the Reynolds stress have been developed. We will consider the second-order closure Reynolds stress models (RSM), the algebraic stress models (ASM) and the nonlinear $k-\varepsilon$ model of Speziale [1988]. These models describe the previous flows better when the $k-\varepsilon$ model fails.

2. RSM: REYNOLDS STRESS MODELS

In order to model different evolutions of the Reynolds components, transport equations for each component R_{ij} have been introduced. Of course, the exact forms of these equations contain unknown correlations which must be modeled in order to obtain a closed system. Moreover, the turbulent energy dissipation appears in the equations. So, one needs to compute ε too. Models employing transport equations for R_{ij} are called second order closure models. The R_{ij} equations can be concisely expressed as (upper cases represent mean flow quantities)

$$\partial_t R_{ij} + U_k \frac{\partial R_{ij}}{\partial x_k} - \frac{\partial D_{ijl}}{\partial x_l} = -P_{ij} + \phi_{ij} + \phi_{ji} + \varepsilon_{ij},$$

where the terms on the left hand side represent respectively the convection and diffusion of $R_{ij} = -\overline{u_i' u_j'}$, and the terms on the right hand side represent the production, pressure redistribution and viscous dissipation.

$$D_{ijl} = \overline{u_i' u_j' u_l'} + \overline{p'(u_i'\delta_{jl} + u_j'\delta_{jl})} + \nu \frac{\partial R_{ij}}{\partial x_l},$$

$$P_{ij} = R_{il}\frac{\partial U_j}{\partial x_l} + R_{jl}\frac{\partial U_i}{\partial x_l},$$

$$\phi_{ij} = -\overline{p' \frac{\partial u_i'}{\partial x_j}},$$

$$\varepsilon_{ij} = 2\nu \overline{\frac{\partial u_i'}{\partial x_l}\frac{\partial u_j'}{\partial x_l}}.$$

Only P_{ij} is exact and does not need any modeling. On the other hand, closure assumptions are necessary for D_{ij}, ϕ_{ij} and ε_{ij}. Several models have been proposed for these terms. For example, Launder et al. [1975] proposed the following transport equation for R_{ij}:

$$\partial_t R + U.\nabla R + c_s' \nabla.(\frac{k}{\varepsilon} R \nabla R)$$

$$= (c'_2 - 1)P - c'_1 \frac{\varepsilon}{k} R - \frac{2c'_2}{3} I \mathrm{tr}(P) - \frac{2}{3}(c'_1 - 1) I \varepsilon,$$

where by definition $P = R(\nabla U + \nabla U^T)$, $k = -\sum R_{ii}/2$, $c'_1 = 1.8$, $c'_2 = 0.6$ and $c'_s = 0.22$. To close the model, we have to compute ε. This will be done using the following transport-diffusion equation where no isotropy assumption is invoked for the diffusion.

$$\partial_t \varepsilon + U \nabla \varepsilon + c_\varepsilon \nabla.(\frac{k}{\varepsilon} R \nabla \varepsilon) = c_1 \frac{\varepsilon}{k} R : \nabla U - c_2 \frac{\varepsilon^2}{k},$$

with $c_\varepsilon = 0.15$, $c_1 = 1.44$ and $c_2 = 1.92$.
The diffusion fluxes have been modeled by simple gradient diffusions and the viscous dissipation is modeled using the Kolmogorov local isotropy of fine scales. This model is valid only at high Reynolds number and does not take into account the presence of a solid wall. The simplest solution to treat the near-wall problem is to use wall-laws. Boundary conditions at a fictitious boundary a small distance away from the wall are used. For instance, knowing the friction velocity at the wall u_τ, we use the following relations to define the value of the variables at $y = \delta$:

$$u = u_\tau (\frac{1}{\chi} \log(\frac{u_\tau \delta}{\nu}) + \beta),$$

$$\overline{u'_1}^2 = 1.2k, \quad \overline{u'_2}^2 = 0.2k, \quad \overline{u'_3}^2 = 0.6k, \quad \overline{u'_1 u'_2} = \overline{u'_2 u'_3} = 0.24k$$

and $\chi = 0.41, \beta = 5.5, c_\mu = 0.09$,

$$k = \frac{u_\tau^2}{\sqrt{c_\mu}}, \quad \varepsilon = \frac{u_\tau^3}{\chi \delta}.$$

A complete description of the model and different type of near-wall techniques is available in Cousteix[1990].

3. INVARIANCE AND REALIZABILITY

As we have seen, invariance requires the models to have the same form in different coordinate systems.
Realizability, as in Shumann [1975], requires from the model to insure positivity of the Reynolds stress components on the diagonal and Schwartz's inequality for all non-diagonal components:

$$\overline{u_i^2} \geq 0, \qquad \overline{u_i u_j} \leq \sqrt{\overline{u_i^2}\ \overline{u_j^2}}, \quad \forall i, j.$$

Because $\overline{u_i u_j}$ is positive semi-definite, it requires also that

$$det(\overline{u_i u_j}) \geq 0.$$

Schumann shows that the exact R_{ij} satisfies the realizability conditions. However, some of the existing second order models are non realizable. On the other hand, realizable R_{ij} models have been proposed but they are quite complicated (Schumann[1975]). Launder's R_{ij} model described above also violates realizability conditions. From a numerical point of view, the positivity of the Reynolds stresses components is important. Special attention must be given to the numerical implementation of such a model to avoid negative values. More precisely, the realizability conditions lead to local stability conditions on the time step for the discretized system. Therefore, at this time such models are only used on simple configurations.

4. ASM: ALGEBRAIC STRESS MODELS

Because of the large number of new transport equations and of severe time step limitations imposed by the realizability conditions, the complete Reynolds stress models always require a big computational effort. To reduce the computational cost, Rodi [1980] proposed an algebraic relation to compute the Reynolds stresses components. Assuming that the transport part of the equations for R_{ij} are proportional to the transport part of the equation for k , we obtain algebraic expressions for R_{ij}. This can be written as

$$R_{ij} = \frac{S_{R_{ij}}}{S_k} k \quad i, j = 1, 3$$

where $S_{R_{ij}}$ and S_k are respectively the right hand side of the R_{ij} and k equations containing different production and destruction terms. More precisely

$$R = -\frac{2}{3} I k - \frac{1 - c_2'}{c_1'} \frac{k}{\varepsilon} (P - \frac{2}{3} I \text{tr}(P)).$$

This is also

$$R(I + \frac{1 - c_2'}{c_1'} \frac{k}{\varepsilon} (\nabla U + \nabla U^T)) = -\frac{2}{3} k I - \frac{4k}{3\varepsilon} I \ \text{tr}(R \nabla U).$$

As these expression contain k and ε, so, k and ε equations must be solved to complete the turbulence model:

$$\partial_t k + U\nabla k + \nabla.(c_k \frac{k}{\varepsilon} R\nabla k) = c_\mu \frac{k^2}{\varepsilon} E - \varepsilon,$$

$$\partial_t \varepsilon + U\nabla \varepsilon + \nabla.(c_\varepsilon \frac{k}{\varepsilon} R\nabla \varepsilon) = c_1 k E - c_2 \frac{\varepsilon^2}{k},$$

where

$$c_1 = 0.1296, \quad c_2 = 1.92, \quad c_\mu = 0.09, \quad c_k = 1, \quad c_\varepsilon = 0.7,$$

and $E = \frac{1}{2}|\nabla U + \nabla U^T|^2$. These equations are different from classical $k - \varepsilon$ by the fact that we have not invoked isotropy for the diffusion terms.

Algebraic stress models are not suitable for flows where the transport of R is important. But they are good when R is driven mainly by the source terms. Therefore, to take account of body forces (buoyancy, rotation), ASM models combine the economy of the two-equation models with the universality of RSM and give predictions similar to the later.

5. NON-LINEAR K-EPSILON MODELS

There are three major problems with Reynolds Stress Models. Firstly, the computational effort is much more since transport equations must be solved for each components of R. Secondly, we have seen that in order to close the R_{ij} equations, closure models have to be introduced and the physical and mathematical signification of the new terms are often uncertain. Finally, the implementation of RSM in existing Navier-Stokes solvers is quite difficult. Realizability deficiencies of these models make this implementation still more tedious.

On the other hand, the $k - \varepsilon$ model is good for attached boundary layers and is easier to incorporate into a Navier-Stokes solver. Therefore, efforts have been made to give a more general form to the $k - \varepsilon$ model to include more general situations. Moreover, we want the new $k - \varepsilon$ model to guarantee the positivity of k and ε, to be frame invariant and also to be unchanged in the limit of 2D turbulence (i.e. The same model should be able to describe 2D and 3D flows).

In fact, we have seen that the major deficiency comes from the Reynolds hypothesis. We have seen in chapter 4 that to be able to take into account three dimensional effects, R should have the following form:

$$R = a\,\mathbf{I} + b\,(\nabla U + \nabla U^T) + c\,(\nabla U + \nabla U^T)^2$$

with a, b and c function of the invariants of $\nabla U+\nabla U^T$, $(\nabla U+\nabla U^T)^2$, k and ε for example.
In this section we describe a so-called *"nonlinear"* $k-\varepsilon$ model. *Nonlinear* means that in these models there is no longer a linear relation between the Newtonian stress and the Reynolds stress tensors. The nonlinear $k-\varepsilon$ we consider is the one proposed by Speziale [1988], Yakhot[1991] where the Reynolds stress tensor is modeled by:

$$R = R_l + R_{nl}$$

where R_l contains the classical expression for the Reynolds stress tensor and R_{nl} the new nonlinear correction:

$$R_l = -\frac{2}{3}k\,\mathbf{I} + 2c_\mu \frac{k^2}{\varepsilon} D$$

with $D = \frac{1}{2}(\nabla U + \nabla U^T)$ and

$$R_{nl} = 4C_D {c_\mu}^2 \frac{k^3}{\varepsilon^2}(D^o - \frac{1}{3}tr(D^o)\,\mathbf{I} + D.D^T - \frac{1}{3}(D:D)\,\mathbf{I})$$

with $C_D = 1.68$ and D^o being the Oldroyd derivative of D:

$$D^o = \partial_t D + U\nabla D - (\nabla U + \nabla U^T)D.$$

The k and ε equations we use to close the model are:

$$\partial_t k + U\nabla k + \nabla.(c_k \frac{k}{\varepsilon} R\nabla k) = R : \nabla U - \varepsilon$$

and

$$\partial_t \varepsilon + U\nabla\varepsilon + \nabla.(c_\varepsilon \frac{k}{\varepsilon} R\nabla\varepsilon) = c_1 \frac{\varepsilon}{k} R : \nabla U - c_2 \frac{\varepsilon^2}{k}$$

where $c_1 = 0.1296$, $c_2 = 1.92$, $c_\mu = 0.09$, $c_k = 1$, $c_\varepsilon = 0.7$. This model has been tested in Speziale[1988] in several configurations where the classical $k-\varepsilon$ fails. For example, this model correctly describes the secondary eddies in a rectangular duct. In the same way, the model gives a better estimation for the length of the recirculation bubble behind a 2D backward step ($7\times$ step height, against 5 for $k-\epsilon$).

6. RNG-BASED K-EPSILON MODEL

Renormalization Group Theory (RNG) has been shown by Yakhot and Orszag [1986] to be a powerful tool for turbulence modeling.

A $k - \epsilon$ model derived by RNG is studied numerically in Martinelli-Yakhot [1989]. For incompressible flows the model is:

$$\partial_t k + u\nabla k - \frac{\nu_T}{2}|\nabla u + \nabla u^T|^2 + \epsilon - \nabla \cdot (\alpha\nu\nabla k) = 0,$$

$$\partial_t \epsilon + u\nabla\epsilon - \frac{\nu_T}{2}\sqrt{\epsilon}C_c Y_\epsilon|\nabla u + \nabla u^T|^2 + \epsilon^{\frac{3}{2}}Y_\epsilon - \nabla \cdot (\alpha\nu\nabla\epsilon) = 0,$$

with $C_c \approx 75$. The eddy viscosity ν_T is related to k^2/ϵ via a differential equation

$$\frac{d}{d\nu_T}\frac{k}{\sqrt{\epsilon}} = 1.72(\nu_T^3 + C_c - 1)^{-\frac{1}{2}}\nu_T.$$

The production term in the ϵ-equation is a function of ν_T given by integrating a differential equation and inverting the result

$$\frac{dY_\epsilon}{d\nu_T} = -0.5764(\nu_T^3 + C_c - 1)^{-\frac{1}{2}}.$$

The function α depends upon ν_T by

$$(\frac{1.3929 - \alpha}{0.3929})^{0.63}(\frac{2.3929 - \alpha}{3.3929})^{0.37} = \nu_T^{-1}$$

The model contains its own low-Reynolds number version and at high Reynolds number it gives the usual law $\nu_T = c_\mu k^2/\epsilon$ but with $c_\mu = 0.084$.

CHAPTER 7

NUMERICAL SIMULATION OF COMPRESSIBLE FLOWS

1. INTRODUCTION

Direct simulation of compressible turbulence is even harder than incompressible turbulence. Present computers are too small to resolve all the scales. Nevertheless we need good Navier-Stokes solvers because turbulence modeling will be done by adding terms in the original equations. The new terms will not destroy the general character of the Navier-Stokes equations, so the turbulent flow solvers will be based upon the Navier-Stokes solvers.

Recall that the general equations for compressible fluids are

$$\partial_t \rho + \nabla.(\rho u) = 0$$

$$\partial_t(\rho u) + \nabla.(\rho u \otimes u) + \nabla p - \mu\Delta u - \frac{1}{3}\mu\nabla(\nabla.u) = f,$$

$$\partial_t[\rho E] + \nabla.[u\rho E] + \nabla.(pu) = f.u + \nabla.\{\kappa\nabla T + [\mu(\nabla u + \nabla u^T) - \frac{2}{3}\mu\mathbf{I}\nabla.u]u\}$$

with the constitutive equation

$$E = \frac{u^2}{2} + T.$$

In dimensionless form μ is the effective viscosity (Re^{-1}), κ the effective thermal diffusivity, f the external volumic forces and γ the adiabatic constant.

With appropriate boundary conditions it defines uniquely, at least for a short time, the velocity field u, the density ρ, and the reduced temperature T. The pressure p is recovered from the perfect gas law

$$p = (\gamma - 1)\rho T.$$

Compressible turbulence is a difficult subject for which some of the general ideas of incompressible turbulence can be used:

- Direct simulations up to a moderate Re;
- Study of singularities, boundary layers, energy spectrum power laws, and in addition shocks.
- Modeling by decomposition into a mean field and a fluctuating field.

There seems to be two clearly different regime:

- When the local Mach number [1] is smaller than 3, the turbulence has similar characteristics as incompressible turbulence
- When the local Mach number is large (>> 3) there is a strong interaction between the shocks and the vortices and a dominance of acoustic phenomena.

Tools to study these problems include experiments but also direct numerical simulation. In this chapter we shall study three numerical methods: a finite volume/finite element method and a full finite element method, both for unstructured meshes. The third method is a finite volume method for structured meshes. The first two can treat any complex boundaries, but they are expensive for transient flows and for simple geometries, an explicit finite difference method or a finite volume like the third method is faster.

2. A FINITE VOLUME / FINITE ELEMENT METHOD

The Navier-Stokes equations can also be written in compact form:

$$\partial_t W + \nabla.F(W) - \nabla.G(W, \nabla W) = 0$$

where

---

[1] The local ach number at x is $|u(x)|/c(x)$ where u and c are the velocity and the speed of sound at x, $c = \sqrt{\gamma p/\rho}$.

$$W = [\rho, \rho u_1, \rho u_2, \rho u_3, \rho E]^T$$

$$F(W) = [F_1(W), F_2(W), F_3(W)]^T$$

$$F_i(W) = [\rho u_i, \rho u_1 u_i + \delta_{1i} p, \rho u_2 u_i + \delta_{2i} p, \rho u_3 u_i + \delta_{3i} p, u_i(E+p)]^T$$

where $p = (\gamma - 1)\rho(E - |u|^2/2)$ and where $G(W, \nabla W)$ is a 2nd order tensor function of W and ∇W such that

$$G_{.,1} = 0, \quad G_{.,2,3,4} = \mu \nabla u + \nabla u^T - \frac{2\mu}{3} I \nabla . u,$$

$$G_{.,5} = \mu(\nabla u + \nabla u^T)u - \frac{2}{3}\mu u \nabla . u + \kappa \nabla (E - \frac{|u|^2}{2})$$

For the numerical simulations we retain the basic idea of performing first a convection step and then a diffusion step.
Consider the model equation

$$\partial_t W + CW + DW = 0$$

discretized semi-implicitly by

$$\frac{1}{\delta t}[W^{n+1} - W^n] + CW^n + DW^{n+1} = 0$$

This scheme is $O(\delta t)$; it is equivalent to

$$\frac{1}{\delta t}[W^{n+1/2} - W^n] + CW^n = 0$$
$$\frac{1}{\delta t}[W^{n+1} - W^{n+1/2}] + DW^{n+1} = 0.$$

Adding both equations proves the equivalence.

2.1 The convection step

So when G is neglected it becomes the Euler equations :

$$\partial_t W + \nabla . F(W) = 0.$$

Upwinding

Upwinding is introduced through the discontinuities of $F(W_h)$ at the inter-element boundaries, either because W_h is a discontinuous approximation of W, or because for all W_h continuous, we know how to associate a discontinuous value at right and at left, at the inter-element edges.

General Framework (Dervieux [1985], Fezoui [1985], Stoufflet et al. [1987]).
For a given triangulation, we associate to each vertex q^i a cell σ^i obtained by dividing each triangle (tetraedra) by the medians (by median planes)(see figure 7.1). Thus we could associate to each piecewise continuous function W_h on a triangulation a P^o function, W_h^p (piecewise constant) in σ^i by the formula

$$W_h^p|_{\sigma^i} = \frac{1}{|\sigma^i|}\int_{\sigma^i} W_h dx.$$

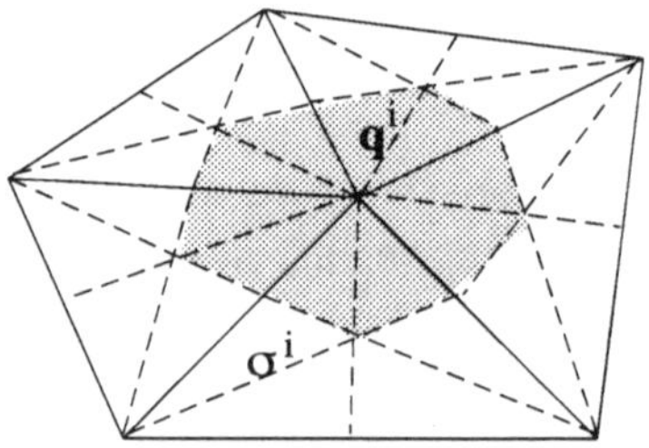

Figure 7.1

By multiplying Euler's equations by the characteristic function of σ^i, after integration (Petrov-Galerkin weak formulation) and after an explicit discretization in time , the following scheme is obtained :

$$W_h^{n+1}(q^i) = W_h^n(q^i) - \frac{k}{|\sigma^i|}\int_{\partial\sigma^i} F_d(W_h^p).n \quad \forall i.$$

An approximation $F_d(W)$ is used instead of $F(W)$ to introduce upwinding in the scheme; $F_d(W)$ is a piecewise constant approximation of $F(W)$ verifying

$$\int_{\partial\sigma^i} F_d(W_h^p).n = \sum_{j\neq i} \Phi(W^p_{h|\sigma^i}, W^p_{h|\sigma^j}) \int_{\partial\sigma^i\cap\sigma^j} n$$

where Φ will be defined as a function of $F(W^p_{h|\sigma^i})$ and $F(W^p_{h|\sigma^j})$. When σ^i touches a wall then the known components of $W_{|\Gamma}$ are used on $\sigma^i \cap \Gamma$ instead of $W_{h|\sigma^j}$.

The *numerical flux function* $\Phi(u, v)$ is chosen according to the qualities sought for the scheme (robustness, precision, ease of programming). In all cases this function should satisfy the consistency relation:

$$\Phi(V, V) = F(V), \text{ for all } V .$$

Definition of the flux Φ
Let $B(W, n) \in R^{5\times 5}$ (4×4 in 2D) be such that

$$F(W)^T n = B^T(W, n)W \quad \forall W, \quad \forall n.$$

Note that F is homogeneous of degree 1 in W, $(F(\lambda W) = \lambda F(W))$ and so $B^T(W, n) = \partial_W F(W)^T n$. It can be shown that B_i is diagonalizable; there exists $S \in R^{5\times 5}$ such that

$$B = S^{-1}\Lambda S$$

where Λ is the diagonal matrix of eigenvalues.
We denote $\Lambda = \{\lambda_i\}_1^5$

$$\lambda_1 = u.n + c, \quad \lambda_2 = u.n - c, \quad \lambda_3 = \lambda_4 = \lambda_5 = u.n$$

and

$$\Lambda^{\pm} = \mathrm{diag}(\pm \max(\pm\lambda_i, 0)), \quad B^{\pm} = S^{-1}\Lambda^{\pm} S,$$

$$|B| = B^+ - B^-, \quad B = B^+ + B^-.$$

We can choose for Φ one of the following formulae :
Steger-Warming[1981] proposed

$$\Phi^{SW}(V^i, V^j) = B^+(V^i)V^i + B^-(V^j)V^j.$$

A more elaborate formula is (Osher[1984])

$$\Phi^{OS}(V^i, V^j) = \frac{1}{2}[F(V^i) + F(V^j) - \int_{V_i}^{V^j} |B(W)| dW]$$

The guiding idea being to get $\Phi_l(V^i, V^j) \cong F(V^i)_l$ if λ_l is positive and $F(V^j)_l$ if $\lambda_l < 0$.
So Φ^{SW} , for instance, can be rewritten as follows :

$$\Phi^{SW}(V^i,V^j)=\frac{1}{2}[F(V^i)+F(V^j)]+\frac{1}{2}[|B(V^i)|V^i-|B(V^j)|V^j],$$

because

$$F(V^i)+F(V^j)=(B^+(V^i)+B^-(V^i))V^i+(B^+(V^j)+B^-(V^j))V^j.$$

The first term, if alone, would yield an upwind approximation. The second, after summation on all the neighbors of i, is an artificial viscosity term. The flux of Osher is built from an integral so that it is C^1 continuous ; the path in R^5 from V^i to V^j is chosen in a precise manner along the characteristics so as to capture exactly singularities like the sonic points (Osher[1984]).

2.3 The diffusion step

The term $\nabla.G(W,\nabla W)$ is a diffusion term when $\rho, T>0$. So a purely explicit scheme combining the convection step for the Euler part of the equations would be

$$W_h^{n+1}(q^i)=W_h^n(q^i)-\frac{k}{|\sigma^i|}\int_{\partial\sigma^i}[F_d(W_h^p)-G(\nabla W_h^n,W_h^n)].n \quad \forall i$$

There is a stability condition on the time step k

$$k\le C\min[\frac{h}{|W_h|},\frac{h^2}{|N|}]$$

where N is a function of κ,μ.

A semi implicit version of the same scheme would be

$$W_h^{n+1}(q^i)=W_h^n(q^i)-\frac{k}{|\sigma^i|}\int_{\partial\sigma^i}[F_d(W_h^p)-G(\nabla W_h^{n+1},W_h^{n+1})].n \quad \forall i$$

The equations at each time step are solved by a relaxation method or a GMRES quasi-Newton method (see Saad[1981]) eventually coupled with a multi-grid strategy.

3. A LEAST SQUARE/SUPG FINITE ELEMENT METHOD

Hughes[1987], Johnson[1987], Mallet[1985] extended SUPG to compressible flows. We present here a version without entropy variables for two dimensional flows. The Navier-Stokes equations are written in the quasi-linear form:

$$\partial_t W + A_i \partial_i W - \partial_i (N_{ij} \partial_j W) = 0 \quad (i, j = 1,, 3)$$

where

$$A_i = \partial_W F_i \quad \textit{and} \quad N_{ij} = (\frac{\partial W}{\partial x_j})^{-1} G_i.$$

Let $\{T_i\}_1^{n_T}$ be a triangulation of Ω; let $\Omega_h = \cup_i T_i$. Let H_h be a space of continuous piecewise polynomial functions on the triangulation defining Ω_h. We define the finite element interpolation space V_h and the test function space V_{0h} by

$$V_{0h} = \{W_h \in [H_h]^5 : \quad BW_h = 0\}$$

and

$$V_h = \{W_h \in [H_h]^5 : \quad BW_h = g_h\}$$

where B and g_h take into account the Dirichlet boundary conditions (recall that all the components of W may not be known at the boundary; B accounts for this problem). The Least square / SUPG finite element method is as follows: find $W_h \in V_h$ such that for all $\Phi_h \in V_{0h}$

$$\int_{\Omega_h} \Phi_h(\frac{1}{\delta t}(W_h^{n+1} - W_h^n) + A_{h_i}^n \partial_i W_h^{n+1}) + \int_{\Omega_h} \partial_i \Phi_h (N_{h_{ij}}^n \partial_j W_h^{n+1})$$
$$+ \int_{\Omega_h} \tau A_{h_k}^{n^T} \partial_k \Phi_h [\frac{1}{\delta t}(W_h^{n+1} - W_h^n) + A_{h_i}^n \partial_i W_h^{n+1} - \partial_i (N_{h_{ij}}^n \partial_i W_h^{n+1})]$$
$$+ \int_{\Omega_h} \delta_{NS}\, \mathbf{I}\, \partial_i \Phi_h \partial_i W_h^{n+1} = 0.$$

Here the first two integrals represent the Galerkin part of the formulation; the third integral is the SUPG stabilization term and the fourth the shock capturing term. Integrals involving derivatives of discontinuous functions are understood as sums of integrals over each elements.
The parameter τ can be defined for instance by an extension of the definition given in $III.7.4:

$$\tau = \frac{h.u}{2||u||(c + ||u||)},$$

where $h = (h_x, h_y)$ is the element length vector and c the local speed of sound.
Another possibility for τ might be the one given in Aliabadi[1993]:

$$\tau = \max_{i=1,2,3}(\frac{h_i|\beta_i|}{2||\beta||(c + |u.\beta|)})$$

where

$$\beta = \frac{\nabla(W^T A_0^{-1} W)^2}{||\nabla(W^T A_0^{-1} W)^2||_0}$$

with $A_0^{-1} = \partial_V W$ where

$$V = (-\frac{E}{T} - s, \frac{u_1}{T}, \frac{u_2}{T}, \frac{u_3}{T}, -\frac{1}{T})^T$$

is a vector of entropy variables and $s = \log(\rho^{1-\gamma} T)$ is the entropy of the system.

The shock-capturing term, δ , is defined in Lebeau[1991] as:

$$\delta_{NS} = \left(\frac{(A_i \partial_i W)^T \tilde{A}_0^{-1} (A_j \partial_j W)}{(\partial_j \xi_l \partial_j W)^T \tilde{A}_0^{-1} (\partial_k \xi_l \partial_k W)} \right)^{1/2}$$

where $\xi_l, l = 1, 2, 3$ are the coordinates in the reference element.
To reduce the numerical diffusion, τ may be set to zero in the regions where the solution is smooth because the numerical viscosity introduced by the SUPG term is sufficient. On the other hand, the numerical viscosity introduced by the shock-capturing term is dominant in regions of discontinuity. Therefore, a better choice for τ is

$$\tau = \max(0, \tau - \frac{\delta}{(c + |u.\beta|)^2})$$

Another optimization consists in removing the SUPG term in regions where the numerical viscosity dominates. For this τ is scaled by

$$f(Pe) = \min(1, \frac{Pe}{3})$$

where $Pe = (\max_i h_i)|u|/\nu$ is the mesh Peclet number.

This formulation is relatively easy to implement and works on unstructured meshes because there is no assumption on the shape of the elements. On the other hand it is more computer intensive than the previous method. It has been generalized by Johnson[1987],Shakib[1988], Hughes et al[1984] to time dependent domains by using a Space-Time finite element.

4. THE PPM FINITE DIFFERENCE METHOD

For direct simulations in simple geometries, it is better to use methods which take advantage of the regularity of the mesh. Colella-Woodward[1984] developed a method which has been used extensively to simulate compressible turbulence in periodic boxes, jets and even wakes (Woodward[1990]).
The Piecewise Parabolic Method (PPM) was originally designed for Euler equations; it uses an explicit semi-lagrangian algorithm with fractional steps; viscous terms can be added in an explicit fashion but it reduces the domain of stability of the method. While the general ideas are easy to understand, the details are involved, so, in view of the fact that we will not use the method with the $k-\epsilon$ model we do not spell out the details. The reader is sent to Collela-Woodward[1984] for these.

Consider the convection equation in one dimension over the real line $R =]-\infty, +\infty[$:

$$\partial_t a + \partial_x(ua) = 0, \quad a(x,0) = a^0(x).$$

If u is constant, the solution $a^{n+1}(x)$ at time $t^{n+1} = t^n + \delta t$ can be deduced from $a^n(x)$ by

$$a^{n+1}(x) = a^n(x - u\delta t).$$

Now we set up the following problem: Given a mesh $\{x_i\}_i$, find the best way of computing a_i^{n+1}, the average of a^{n+1} on $]x_i, x_{i+1}[$, from the knowledge of a_i^n only?

In PPM this is done by constructing a piecewise parabolic approximation a_h^n of a^n with means on $]x_i, x_{i+1}[$ equal to a_i^n, and such that a_h^n does not have more extrema than a^n in each interval. Then a_h^{n+1} is computed by convection, i.e. $a_h^{n+1}(x) = a_h^n(x - u\delta t)$ and the means of the result are calculated so that the process can be reiterated at the next time step.

To extend the method to Euler equations in one space dimension it suffices to replace the convection step by a Riemann solver (see for example Lax et al[1979], Godlewski-Raviart[1992]). The method can be extended to several space dimensions also by performing every time step several fractional one dimensional steps; to preserve second order accuracy a scheme of the type $x_1 - x_2 - x_3 - x_3 - x_2 - x_1$ must be used (Strang[1968], Beale-Majda[1981]).
Several additional features of PPM contribute to its effectiveness. A contact discontinuity detection algorithm steepens the parabolic interpolation in zones near such discontinuities. Shock detection is also performed; because of the sharpness in shock structures in PPM, numerical noise can be generated in certain pathological cases. Accordingly, localized numerical diffusion is turned on in zones containing sharp shock jumps.
The method has a stability condition $\delta t < u/\delta x$.

Even with Euler equations, PPM is able to simulate turbulence and viscous effects because of its numerical dissipation. An estimate of the local effective viscosity ν_{eff} is given in Woodward[1991]

$$\nu_{eff} = 0.462c(M + 0.25)^{3.15}L^{-4}\delta x^3.$$

where c is the local speed of sound, M the mach number of the flow, L a characteristic length of the macro-structures and δx the mesh size.

5. NUMERICAL RESULTS

In this section we present some numerical results obtained with the methods detailed above.

Compressible mixing layer
Figure 7.2 shows a fairly intensive two dimensional transient computation with a Finite Volume / Finite Volume method in the same family as the one described in section 2. It simulates turbulence in a mixing layer. The computational domain is a box of size 80. Periodic boundary conditions are applied horizontally and Neumann conditions vertically. The initial conditions are

$$u_1 = \frac{1}{2}\tanh(2x_2) \quad u_2 = 0$$

$$T = 1 + \frac{\gamma - 1}{8}M^2(1 - (2u_1)^2)$$

$$\rho = \frac{1}{T}$$

A small random solenoidal perturbation is added to it. The mesh is 201×201.

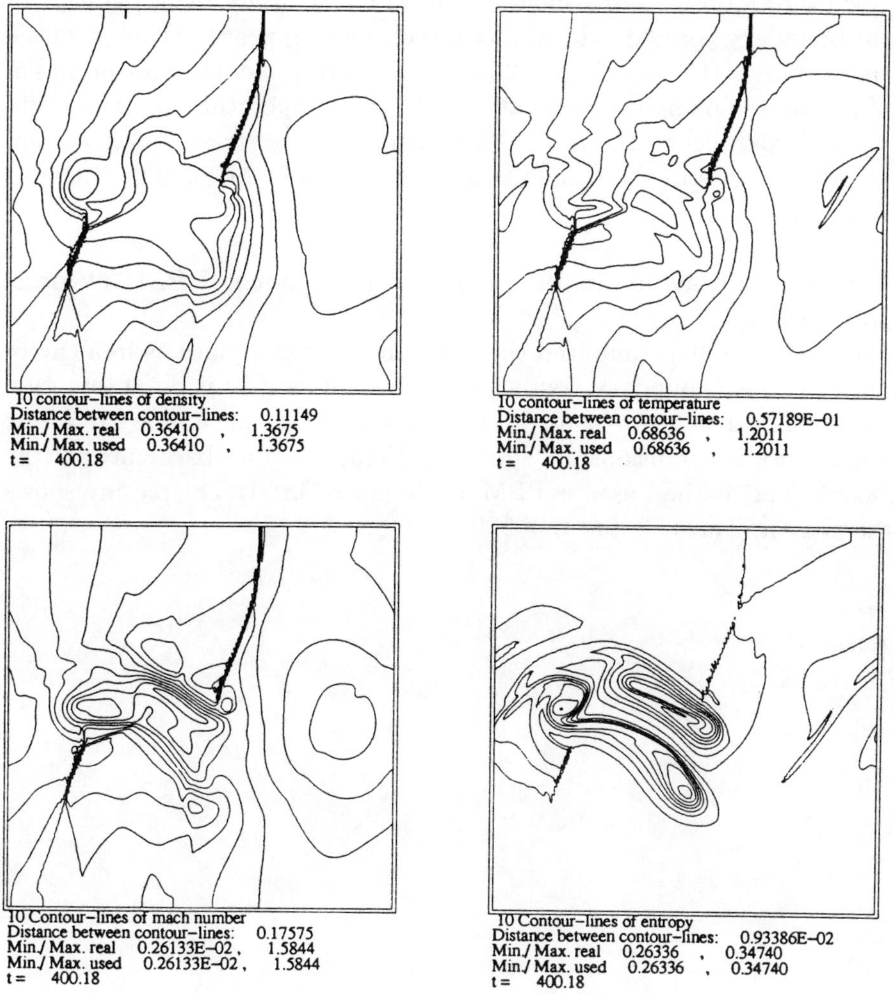

Figure 7.2: Flow in a mixing layer (Guillard[1992]). The flow in a mixing layer is computed with a Finite Volume / Finite Element method similar to the one of §2. It shows density, temperature, Mach and entropy level lines at time 400 for an initial flow with Mach=0.8, Re=1000, Pr=0.7

Direct simulation over the canopy of Hermes

Figure 7.3 (see back cover of the book) shows an intensive three dimensional computation of the Navier-Stokes equations with the SUPG / Least square method described in section 3 but using entropy variables. This computation was performed at Dassault Aviation. It is an external flow over an obstacle (Hasholder et al[1992]) to study the separation of the boundary layers due to surface discontinuities near the canopy of the space shuttle Hermes. $M_\infty = 20, \rho_\infty, T_\infty$ corresponding to air at 60 km of altitude. The figure shows the mesh, the temperature and the white lines are parallel to the normal component of the stress tensor. There are 10^5 vertices in the triangulation. (Courtesy of M. Mallet, Dassault Aviation).

Direct simulation in a periodic box with PPM (Woodward[1991])

Edgar et al [1991] computed the solution of Euler equations in a cavity with periodic boundary conditions. The flow is driven by an external force. The results of a direct simulation of turbulent flow in a box with periodic conditions are shown on Figure 7.4 (see back cover of the book). The method used is PPM, as described in §1. The picture shows vorticity at Mach=5 and Re=500.

CHAPTER 8

COMPRESSIBLE REYNOLDS EQUATIONS

1. INTRODUCTION

In this chapter we shall derive an averaged compressible Navier-Stokes equations. Our aim is to have, as in the incompressible case, an averaged form as close as possible to the original form of the equations. For this reason we will introduce a new averaging operator, called the Favre average, or density weighted average. Indeed, in the compressible case, the Reynolds average, classically used for incompressible flows, leads to several new terms which are hard to model. First recall the general equations for compressible fluids in adimensional form:

$$\partial_t \rho + \nabla.(\rho u) = 0,$$

$$\partial_t(\rho u) + \nabla.(\rho u \otimes u) = \nabla.\tau,$$

$$\partial_t(\rho E) + \nabla.(u \rho E) = \nabla.(\tau u) + \kappa \Delta T,$$

where $\tau = -p\ \mathbf{I} + \mu(\nabla u + \nabla u^t) - \frac{2}{3}\mu \nabla.u\ \mathbf{I}$ is the Newtonian stress tensor, μ the effective viscosity, κ the effective thermal diffusivity, $E = |u|^2/2 + T$ the total energy. With appropriate boundary conditions, this system defines the velocity field u, the density ρ and the reduced temperature T.

2. THE FAVRE AVERAGE

Let $\bar{\rho}$ and $\bar{u}$ denote the mean parts of ρ and u and let ρ' and u' be the fluctuations. Then

$$\rho = \bar{\rho} + \rho' \quad \text{with} \quad \bar{\rho}' = 0,$$

$$u = \bar{u} + u' \quad \text{with} \quad \bar{u}' = 0.$$

For term like ρu we will have

$$\rho u = \bar{\rho}\bar{u} + \bar{\rho}u' + \rho'\bar{u} + \rho' u'.$$

And so

$$\overline{\rho u} = \bar{\rho}\bar{u} + \overline{\rho' u'}.$$

This makes the Reynolds average difficult to use in compressible situations. To avoid this problem Favre [1976] proposed a new density weighted average operator.

2.1. Definition

Assume that the filter, denoted $^{-}$, satisfies the properties of chapter 4, namely:

• It is a linear operator. • It commutes with time and space derivates.

• It is idempotent in the sense that $\overline{\bar{f} g} = \bar{f}\bar{g}$ for all g.

Let a be a fluctuating function and let $\bar{a}$ be the filtered part. The oscillating part being a', we have

$$a = \bar{a} + a'.$$

Now consider the density weighted average $\tilde{a}$:

Definition

Favre average is

$$\tilde{a} = \frac{\overline{\rho a}}{\bar{\rho}}$$

Favre fluctuating part is

$$a'' = a - \tilde{a}$$

Proposition

Favre filter is a linear operator; it is idempotent in the sense that

$$\widetilde{f\tilde{g}} = \tilde{f}\tilde{g}, \quad \overline{f\tilde{g}} = \overline{\tilde{f} g} = \bar{f}\tilde{g}.$$

However it does not commute with derivates.

Remark
Notice that

$$\tilde{\tilde{a}} = \frac{1}{\bar{\rho}}\overline{\rho(\frac{\overline{\rho a}}{\bar{\rho}})} = \tilde{a}, \quad \tilde{a''} = 0,$$

and

$$\overline{\rho a} = \bar{\rho}\tilde{a}, \quad \overline{\rho a''} = 0.$$

Proof of Proposition

$$\widetilde{f\tilde{g}} = \frac{1}{\bar{\rho}}\overline{\rho f \frac{1}{\bar{\rho}}\overline{\rho g}} = \frac{1}{\bar{\rho}}\overline{\rho f}\frac{1}{\bar{\rho}}\overline{\rho g} = \tilde{f}\tilde{g}$$

The rest is proved in a similar way.

2.2. Averaging the Continuity Equation

The equation

$$\partial_t \rho + \nabla.(\rho u) = 0$$

becomes

$$\partial_t(\bar{\rho} + \rho') + \nabla.[(\bar{\rho} + \rho')(\bar{u} + u')] = 0.$$

Apply the filter and get

$$\partial_t \bar{\rho} + \nabla.(\overline{\rho u}) = 0.$$

But, we have

$$\overline{\rho u} = \bar{\rho}\frac{\overline{\rho u}}{\bar{\rho}} = \bar{\rho}\tilde{u}.$$

Therefore,

$$\partial_t \bar{\rho} + \nabla.(\bar{\rho}\tilde{u}) = 0.$$

So, the Favre average enables us to keep the original form for the density equation.

2.3. Averaging the Momentum Equation

For the momentum equation

$$\partial_t(\rho u) + \nabla.(\rho u \otimes u) - \nabla.\tau = 0$$

where

$$\tau = -p\mathbf{I} + \mu(\nabla u + \nabla u^T) - \frac{2}{3}\mu(\nabla . u)\mathbf{I}.$$

We have:

$$\overline{\rho u} = \bar{\rho}\tilde{u},$$

and

$$\begin{aligned}\overline{\rho u \otimes u} &= \overline{\rho(u'' + \tilde{u}) \otimes (u'' + \tilde{u})} \\ &= \bar{\rho}\tilde{u} \otimes \tilde{u} + \overline{\rho u'' \otimes u''}.\end{aligned}$$

Because for any functions a, b:

$$\overline{\rho \tilde{a} b''} = \overline{\rho \frac{\overline{\rho a}}{\bar{\rho}} b''} = \frac{\overline{\rho a}}{\bar{\rho}} \overline{\rho b''} = \bar{\rho}\tilde{a}\widetilde{b''} = 0.$$

Therefore,

$$\partial_t(\bar{\rho}\tilde{u}) + \nabla.(\bar{\rho}\tilde{u} \otimes \tilde{u}) - \nabla.(\bar{\tau}) = -\nabla.(\overline{\rho u'' \otimes u''})$$

with

$$\bar{\tau} = -\bar{p}\mathbf{I} + \mu(\nabla\bar{u} + \nabla\bar{u}^T) - \frac{2}{3}\mu\nabla.\bar{u}\mathbf{I}.$$

Definition
The Reynolds Stress tensor is defined as

$$R_{ij} = -\overline{\rho u_i'' u_j''} = -\bar{\rho}\widetilde{u_i'' u_j''}.$$

2.4. Averaging the Energy Equation

Recall the energy equation:

$$\partial_t(\rho E) + \nabla.(u\rho E) = \nabla.\{\kappa\nabla T + \tau u\}.$$

We have:

$$\overline{\rho E} = \bar{\rho}\tilde{E}$$

and

$$\overline{u\rho E} = \overline{\rho(u'' + \tilde{u})(E'' + \tilde{E})} = \bar{\rho}\tilde{u}\tilde{E} + \overline{\rho u'' E''}.$$

So, the averaged energy equation proposed, is

$$\partial_t(\overline{\rho}\tilde{E}) + \nabla.(\overline{\rho}\tilde{u}\tilde{E}) = -\nabla.(\overline{\rho u'' E''}) + \nabla.(\overline{\tau u}) + \nabla.(\kappa\nabla\overline{T}).$$

Now the constitutive law must be averaged too:

$$E = \frac{|u|^2}{2} + T.$$

So

$$\tilde{E} = \frac{|\tilde{u}|^2}{2} + \tilde{T} + \widetilde{\tilde{u}.u''} + \frac{1}{2}\widetilde{|u''|^2} = \frac{|\tilde{u}|^2}{2} + \tilde{T} + k,$$

where k is defined as

Definition
The turbulent kinetic energy is

$$k = \frac{1}{2}|\widetilde{u''^2}|.$$

Similarly the law of perfect gas being $p = (\gamma - 1)\rho T$, it averages into

$$\overline{p} = (\gamma - 1)\overline{\rho}\tilde{T}.$$

Now we proceed to give an alternative form for the averaged Energy equation.
We have

$$E'' = \frac{1}{2}(|\tilde{u} + u''|^2 - |\tilde{u}|^2) + T'' = \frac{1}{2}|u''|^2 + \tilde{u}\cdot u'' + T''.$$

So

$$\overline{\rho u'' E''} = \overline{\rho u'' T''} + \overline{\rho u'' \otimes u''}\tilde{u} + \frac{1}{2}\overline{\rho |u''|^2 u''}.$$

With Vandromme[1984] assume that (H1) $\overline{\rho u_i'' u_j'' u_k''}$ is small.
Then

$$\overline{\rho u'' E''} = \overline{\rho u'' T''} + \overline{\rho u'' \otimes u''}\tilde{u}$$

To estimate $\overline{\tau u}$ we notice that

$$\overline{\tau u} = -(\gamma - 1)\overline{\rho}\tilde{T}\tilde{u} - (\gamma - 1)\overline{\rho T'' u''} + \overline{[\mu(\nabla u + \nabla u^T) - \frac{2}{3}\mu\nabla \cdot u I]u}.$$

We shall model the last terms by simply replacing bars by tildes, even in products, on the ground that in practice μ is very small anyway.

(H2) Assume that

$$\overline{\mu(\nabla u+\nabla u^T)u}-\frac{2}{3}\overline{\mu u\nabla\cdot u}+\nabla.(\kappa\nabla\overline{T})$$
$$\simeq \mu\tilde{u}(\nabla\tilde{u}+\nabla\tilde{u}^T)-\frac{2}{3}\mu\tilde{u}\nabla\cdot\tilde{u}+\nabla.(\kappa\nabla\tilde{T})$$

Then the averaged Energy equation is now

$$\partial_t(\overline{\rho}\tilde{E})+\nabla.(\overline{\rho}\tilde{u}\tilde{E})+\gamma\nabla.(\overline{\rho u''T''})+\nabla.(R\tilde{u})=\nabla.(\tilde{\tau}\tilde{u})+\nabla.(\kappa\nabla\tilde{T}).$$

2.5. The Averaged Navier-Stokes Equations

Collecting all terms in the previous analysis gives:

$$\partial_t\overline{\rho}+\nabla.(\overline{\rho}\tilde{u})=0,$$
$$\partial_t(\overline{\rho}\tilde{u})+\nabla.(\overline{\rho}\tilde{u}\otimes\tilde{u})+\nabla\overline{p}$$
$$=\nabla.[\mu(\nabla\bar{u}+\nabla\bar{u}^T)-\frac{2}{3}\mu\nabla.\bar{u}\mathbf{I}]+\nabla.R$$
$$\partial_t(\overline{\rho}\tilde{E})+\nabla.(\overline{\rho}\tilde{u}\tilde{E})=\nabla.(\gamma\overline{\rho u''T''}+\kappa\nabla\tilde{T})-\nabla.((R+\overline{p}I)\tilde{u})$$
$$+[\mu(\nabla\tilde{u}+\nabla\tilde{u}^T)-\frac{2}{3}\mu\nabla\cdot\tilde{u}I]\tilde{u}$$

and $\bar{p}=(\gamma-1)\bar{\rho}\tilde{T}$, $\tilde{E}=\frac{1}{2}|\tilde{u}|^2+\tilde{T}+k$.
To close this system, it is necessary to evaluate or to model the Reynolds stresses $R_{ij}=-\bar{\rho}\widetilde{u_i''u_j''}=-\overline{\rho u_j''u_i''}$, the turbulent heat fluxe $\bar{\rho}\widetilde{T''u''}=\overline{\rho T''u''}$ and the turbulent kinetic energy $k=\frac{1}{2}\widetilde{|u"|^2}$.

2.6 Reynolds Equations

The derivation of the Reynolds equations for compressible flows is somehow similar to the incompressible case. The details of the derivation are however long and tedious. We will just give here the more important steps and the complete equations. Details can be found in Vandromme [1984].
Consider the following equations obtained by multiplying the j and k components of the non averaged momentum equation by $u"_k$ and $u"_j$ respectively:

$$u"_k[\partial_t(\rho u_j)+\partial_i(\rho u_i u_j)-\partial_i\tau_{ij}]=0$$

$$u"_j[\partial_t(\rho u_k)+\partial_i(\rho u_i u_k)-\partial_i\tau_{ik}]=0.$$

Now apply the filter to the sum. Because ρ can be pulled in and out of the derivatives,we obtain

$$\partial_t(\bar{\rho}\widetilde{u"_j u"_k}) + \partial_i(\bar{\rho}\tilde{u}_i\widetilde{u"_j u"_k})$$
$$= -\partial_i D_{ijk} + P_{jk} + P_{kj} + \phi_{jk} + \phi_{kj} + W_{jk} + W_{kj} - \bar{\rho}\varepsilon_{jk} - \bar{\rho}\varepsilon_{kj}.$$

The physical meaning of the terms in the right hand side is similar to the incompressible case. The first term is a complex diffusion effect which includes different contributions:

$$D_{ijk} = [\bar{\rho}u"_i\widetilde{u"_j u"_k} + \delta_{ij}\overline{u"_k p'} + \delta_{ik}\overline{u"_j p'} - \mu\overline{\tilde{S}_{ij}u"_k} - \mu\overline{\tilde{S}_{ik}u"_j}],$$

where

$$S = \nabla u + \nabla u^T - \frac{2}{3}\nabla.uI.$$

Two terms represent production by the mean flow:

$$P_{jk} = -\bar{\rho}\widetilde{u"_j u"_i}\frac{\partial\tilde{u}_k}{\partial x_i}.$$

Two other terms include effects of velocity fluctuations on the mean pressure:

$$\phi_{kj} = -\overline{u"_k}\partial_i\bar{p}.$$

In the same way two terms take into account interactions between pressure and velocity fluctuations

$$W_{jk} = -\overline{\partial_j p' u"_k} \simeq \overline{p'\partial_j u"_k}.$$

Finally, two terms represent the dissipation by viscosity:

$$\bar{\rho}\varepsilon_{jk} = -\overline{\mu\partial_i S"_{ij}u"_k} \simeq \overline{\mu S"_{ij}\partial_i u"_k}.$$

(H3) Erogidicity and integration by parts lead to the approximtion of the terms above, as for incompressible fluids.
These equations are not closed because new correlations have appeared. Therefore, the new terms have to be modeled.

3. FURTHER MODELING

3.1. The k Equation for Compressible Flows

From the previous equations for R_{ij} we can derive an equation for the turbulent kinetic energy k using its definition:

$$k = \frac{1}{2} \sum_{i=1,3} \widetilde{u_i''^2}.$$

From Reynolds' equations above, it is found that

$$\partial_t(\bar{\rho}k) + \nabla \cdot (\bar{\rho}\tilde{u}k) = -\nabla \cdot D + P + \phi + W - \bar{\rho}\varepsilon$$

where

$$D_i = \frac{1}{2}\bar{\rho}\widetilde{u''_i u''_j u''_j} + \delta_{ij}\overline{u''_j p'} - \mu\tilde{S}_{ij}\overline{u''_j} \quad \text{(Diffusion)}$$

$$P = -\bar{\rho}\widetilde{u''_j u''_i}\partial_i\tilde{u_j} \quad \text{(Production)}$$

$$\phi = -\overline{u''} \cdot \nabla\bar{p} \quad \text{(Mean Velocity-Pressure interaction)}$$

$$W = \overline{p'\nabla \cdot u''} \quad \text{(Velocity-Pressure fluctuations interaction)}$$

$$\bar{\rho}\varepsilon = \overline{\mu S''_{ij}\frac{\partial u''_j}{\partial x_i}} = \frac{\mu}{2}\overline{|\nabla u'' + \nabla u''^T|^2} - \frac{2}{3}\mu\overline{|\nabla . u''|^2} \quad \text{(Dissipation).}$$

This equation is similar to the one obtained for incompressible flows. Some new terms taking into account the compressibility effects appear and the definition of ε is slightly different.

Definition
The rate of turbulent energy dissipation, ε, is defined by

$$\bar{\rho}\varepsilon = \frac{\mu}{2}\overline{|\nabla u'' + \nabla u''^T|^2} - \frac{2}{3}\mu\overline{|\nabla . u''|^2}$$

3.2 The ε Equation for Compressible Flows

The derivation of an equation for ε for compressible flows is much more difficult than for incompressible case. We will not give here an exact account for the ε-equation (see Vandromme[1984]). The final equation before modeling is

$$\partial_t\bar{\rho}\varepsilon + \partial_i\bar{\rho}\tilde{u_i}\varepsilon = F$$

where the right hand side F has more than 20 terms !

3.3. Modeling the Reynolds Stress and the Heat flux

(H4) To model the Reynolds stress tensor R, we suppose it to be a function of the deformation tensor.

Then by a frame invariance argument (see Chapter 3) in two dimensions the only possible form is:

$$R = -\frac{2}{3}\bar{\rho}k\mathbf{I} + \mu_T(\nabla\tilde{u} + \nabla\tilde{u}^T - \frac{2}{3}\nabla.\tilde{u}\mathbf{I}),$$

with the turbulent viscosity given by

$$(H5) \qquad \mu_T = c_\mu \bar{\rho}\frac{k^2}{\varepsilon}.$$

Recall here again that this model for R is quite general in 2D but somewhat restrictive in 3D where the general form should take into account all the invariants of $\nabla\tilde{u} + \nabla\tilde{u}^T - \frac{2}{3}\nabla.\tilde{u}\mathbf{I}$ and be

$$R = -\frac{2}{3}\bar{\rho}k\mathbf{I} + \mu_T(\nabla\tilde{u} + \nabla\tilde{u}^T - \frac{2}{3}\nabla.\tilde{u}\mathbf{I}) + \alpha_T(\nabla\tilde{u} + \nabla\tilde{u}^T - \frac{2}{3}\nabla.\tilde{u}\mathbf{I})^2.$$

Similarly the turbulent heat flux $-\overline{\rho T''u''}$ is modeled by:

$$(H5) \qquad -\gamma\overline{\rho T''u''} = \kappa_T\nabla\tilde{T} \text{ with } \kappa_T = \frac{\gamma}{Pr_T}\mu_T$$

where the turbulent Prandtl number is $Pr_T = 0.9$.

3.4. Closing the k Equation

Consider the equation for k obtained in Section 3.1:

$$\partial_t(\bar{\rho}k) + \nabla\cdot(\bar{\rho}\tilde{u}k) = \nabla\cdot D + P + \phi + W - \bar{\rho}\varepsilon.$$

The production term P is known once R is known. We model the unknown terms as follows :

(H6) The turbulent diffusion term $\nabla\cdot D$ is supposed to have the same structure as the viscous one for high Reynolds number flows:

$$(H7) \qquad \nabla\cdot D = \nabla\cdot((\mu + \frac{\mu_T}{\sigma_k})\nabla k).$$

σ_k is called the turbulent Prandtl-Schmidt number of k,
(H8) The pressure dilatation effect can be modeled following Grasso [1989] :

$$\phi + W = -\overline{u"}\nabla\bar{p} + \overline{p'\nabla u"} = \frac{c_k\bar{\rho}}{\bar{p}}(\tilde{u}\cdot R\tilde{u})\nabla\cdot\tilde{u}.$$

This hypothesis is reasonable only for statistically steady flows. Hence, the model can only be used for such a flow. Furthermore, the explicit dependence on the mean velocity $\tilde{u}$ in this model makes it clearly non Galilean invariant. In addition, new terms have been introduced to take into account possible density variations (see Vandromme[1984]) such as multi component fluids with or without combustion. All these terms introduce new constants which must be numerically tuned. In its simplest form the k equation is

$$\partial_t\bar{\rho}k + \nabla\cdot(\bar{\rho}\tilde{u}k) - \nabla\cdot((\mu + \frac{\mu_T}{\sigma_k})\nabla k) = R : \nabla\tilde{u} - \bar{\rho}\varepsilon + \frac{c_k\bar{\rho}}{\bar{p}}(\tilde{u}\cdot R\tilde{u})\nabla\cdot\tilde{u}.$$

Remark

When ρ is constant and $\nabla.u = 0$ the equation for k is the one found in the case incompressible flows.

3.5 The modeled ε equation

As we have seen the exact equation for ε in the compressible case is too complicated to be a good starting point for a modeling process. Let

$$\theta = k/\varepsilon.$$

The dimension of θ is that of a time. Therefore one should be able to model an equation for ε by starting from the k equation and rescaling it with a time (ϵ is k scaled by θ). The numerical coefficients of the equation will have to be changed also.

(H9) The ε equation proposed is

$$\partial_t\bar{\rho}\varepsilon + \nabla\cdot(\bar{\rho}\tilde{u}\varepsilon) - \nabla\cdot((\mu + \frac{\mu_T}{\sigma_\varepsilon})\nabla\varepsilon) =$$
$$\frac{\varepsilon}{k}(c_1 R : \nabla\tilde{u} - c_2\bar{\rho}\varepsilon + c_3 c_k\frac{\bar{\rho}}{\bar{p}}\tilde{u}\cdot R\tilde{u}\nabla\cdot\tilde{u}).$$

As the k equation is only valid in high Reynolds number regions, the ε equation cannot do better.

The constants $c_\mu, c_\epsilon, c_{1\epsilon}, c_{2\epsilon}$ are the same as in the incompressible case because the model must be the same in this case. The new constant $c_{3\epsilon}$

is chosen to enable the model to reproduce the correct level of turbulence generated downstream of shocks (Grasso[1989]). Thus

$$\sigma_k = 1,\ \sigma_\epsilon = 1.3,\ c_\mu = 0.09,\ c_\epsilon = 1.3,\ c_{1\epsilon} = 1.45,\ c_{2\epsilon} = 1.92,\ c_{3\epsilon} = 2.$$

4. LOW REYNOLDS NUMBER REGIONS

The previous model has been established under the hypothesis that the Reynolds number is high; so we cannot consider regions close to a solid wall where the local Reynolds number decreases to zero.

4.1 Wall Laws

The most widely used approach when dealing with the near-wall regions is to avoid solving the Navier-Stokes equations (including the turbulence model) right up to the wall. As for incompressible fluids the boundary Σ of the computational domain is placed at a small distance δ away from the wall in the high Reynolds number region (fully developed turbulent region). Empirical laws as those used for incompressible fluids are then used to define boundary conditions on Σ. Indeed, the idea is that near walls, the flow is subsonic so matching with a log profile should be reasonable.

In the following, the subscript w and δ mean at the wall and at Σ. The friction velocity which is by definition $u_\tau = (\frac{\mu}{\rho}|\frac{\partial u}{\partial n}|)^{1/2}$ is computed by

$$u_\tau = \begin{cases} U_c/y^+ & \text{if } y^+ \leq 20 \\ U_c/(\frac{1}{\chi}\log(y^+) + \beta) & \text{if } 20 \leq y^+ \leq 100, \end{cases}$$

where $\chi = 0.41$, $\beta = 5.5$, $y^+ = \rho u_\tau y/\mu$ and U_c is given by the Van Driest compressibility transformation Gorski[1986]:

$$U_c = \int_0^{u_\delta} (\frac{\rho_\delta}{\rho_\omega})^{1/2} du \approx u_\delta (\frac{\rho_\delta}{\rho_w})^{\frac{1}{2}}.$$

The boundary condition for ρ is obtained by writing that $\frac{\partial p}{\partial n} = 0$ and using $p = (\gamma - 1)\rho T$. This gives

$$0 = \rho\frac{\partial T}{\partial n} + T\frac{\partial \rho}{\partial n} \approx T\frac{\rho_\delta - \rho_w}{\delta} + \rho_\delta(\frac{T_\delta - T_w}{\delta}).$$

This formula determines ρ_w because the rest is known. Then the boundary conditions for k and ε are :

$$k_\delta = \frac{u_\tau^2}{\sqrt{c_\mu}}\frac{\rho_w}{\rho_\delta}, \quad \varepsilon_\delta = \frac{u_\tau^3}{\chi\delta}(\frac{\rho_w}{\rho_\delta})^{\frac{3}{2}}.$$

Let $S = (\mu + \mu_T)(\nabla u + \nabla u^T - \frac{2}{3}\nabla \cdot uI)$ be the viscous part of the stress tensor, s a unit vector tangent to S and n the unit outward normal to Γ, the boundary conditions on the fluid variables are :

$$u \cdot n = 0,$$

$$s^T Sn = -\rho u_\tau^2,$$

$$u^T Sn + (\kappa + \kappa_T)\frac{\partial T}{\partial n} = -\rho u_\tau^2 (u \cdot s) + \gamma \rho T_\tau u_\tau,$$

where T_τ is the friction temperature defined by :

$$T_\tau = \frac{\kappa + \kappa_T}{u_\tau}\frac{\partial T}{\partial n}.$$

Note that u_τ is a function of $u_\delta = u \cdot s$. To compute T_τ in the fully turbulent region, we assume that, after translation and symmetry, the temperature profile is similar to the velocity one, therefore T_τ is evaluated (see Brun[1988]) as

$$T_\tau = \begin{cases} \frac{T_w - T_\delta}{Pr\ y^+} & \text{if } y^+ \leq 20 \\ \frac{T_w - T_\delta}{\frac{1}{\chi}\log(y^+) + c} & \text{if } y^+ \geq 20 \end{cases}$$

For a cold wall$Pr = 0.72$, $c = 2$.

4.2 A modified wall law for ε

In the previous section the friction velocity u_τ is computed from the mean flow. It seems interesting to introduce a more appropriate velocity scale, $\sqrt{k}$, dedicated to describe turbulent quantities.

If S is placed at a distance δ, from the wall, in the fully turbulent region where we assume equilibrium between production and dissipation of turbulent kinetic energy, then the following holds:

$$-\overline{\rho u"v"}\frac{\partial \bar{u}}{\partial y} = \bar{\rho}\varepsilon.$$

Now, define

$$u_\tau^k = c_\mu^{\frac{1}{4}} k^{\frac{1}{2}} \quad \text{and} \quad u_\tau^l = \frac{u_T}{(1/\chi)\log y^+ \beta},$$

These two friction velocities are such that $\overline{-\rho u"v"} = \bar{\rho}_w u_\tau^{k^2}$ and $\partial_y \bar{u} = u_\tau^l/\chi\delta$. Then the boundary condition on ε is set to be

$$\varepsilon = (\frac{\rho_w}{\rho_\delta})^{\frac{3}{2}}({u_\tau^k}^2 u_\tau^l/\chi\delta).$$

This approach improves the prediction of the turbulent quantities in regions where the friction at the wall decreases but where the turbulence is still high (Laurence[1990]) .

4.3 Low-Reynolds Models

An alternative to wall-laws might be to modify the $k-\varepsilon$ equations to take into account the damping at the wall. The low-Reynolds $k-\varepsilon$ models are the same as for incompressible flows (see chapter 4).

4.4 Matching with an algebraic model

Let us illustrate this approach with a simple algebraic viscosity model near a wall.

With a Baldwin-Lomax or Cebeci-Smith model, μ_T is computed by an algebraic expression in terms of the distance to the wall y and the mean velocity gradient

$$\mu_T = \mu_T(y, |\nabla u|).$$

The formula is somewhat involved so it is given in Appendix A8. Then k and ε are given by

$$k = \frac{\mu_T}{c_\mu \rho l_\mu} \quad \text{and} \quad \varepsilon = \frac{k^{3/2}}{l_\varepsilon}$$

where l_μ and l_ϵ are two length scales

$$l_\mu = \chi {c_\mu}^{-3/4} y(1 - e^{-\frac{y^+}{70}})$$

$$l_\epsilon = \chi {c_\mu}^{-3/4} y(1 - e^{-\frac{y^+}{2\chi}{c_\mu}^{3/4}})$$

with $y^+ = \frac{u_\tau \rho}{\mu} y$.

4.5 Matching with a one equation model

An alternative is to use a one equation turbulence model exactly as in the incompressible case (see §IV.4). Then in the low-Reynolds regions the $k-$ equation is unchanged but the $\varepsilon-$equation is replaced by the formula $\varepsilon = k^{3/2}/l_\varepsilon$ and an algebraic expression is used for l_ε (as above) in terms of y and ∇u. In Mohammadi [1992] comparisons can be found between results obtained by a two-layers approach and the previous algebraic model for a supersonic flow over a compression ramp. This two-layers approach seems to be more general and easier to implement than the previous two layers method with an algebraic model because there is no need for special geometrical considerations involving the distance from a wall.

SUMMARY

The $k-\varepsilon$ model for compressible flows is summarized below. Some terms are underlined; these may be neglected to obtained what will be refered in the sequel as a reduced model. Bars and tildes are removed for clarity; thus in reality ρ stands for $\bar{\rho}$, u for $\tilde{u}$, p for $\bar{p}$, E for $\tilde{E}$ and T for $\tilde{T}$.

$$\begin{aligned}
&\partial_t\rho+\nabla.(\rho u)=0,\\
&\partial_t(\rho u)+\nabla.(\rho u\otimes u)+\nabla p\\
&\qquad=\nabla.[\mu(\nabla u+\nabla u^T)-\frac{2}{3}\mu\nabla.u\mathbf{I}]+\nabla.R,\\
&\partial_t(\rho E)+\nabla.(\rho uE)=\nabla.((\kappa+\kappa_T)\nabla T+Ru)+\\
&\qquad+\nabla.\{-pu+[\mu(\nabla u+\nabla u^T)-\frac{2}{3}\mu\nabla\cdot uI]u\}.
\end{aligned}$$

Constitutive relations are modified

$$E=\frac{u^2}{2}+T+\underline{k},\quad p=\rho T(\gamma-1).$$

The Reynolds tensor is modeled by

$$R=-\frac{2}{3}\underline{\rho k}\mathbf{I}+\mu_T[(\nabla u+\nabla u^T)-\frac{2}{3}\nabla.u\mathbf{I}]$$

with $\mu_T=c_\mu\rho\frac{k^2}{\varepsilon}$, $\kappa_T=\frac{\gamma}{Pr_T}\mu_T$, $Pr_T=0.9$ and two equations for k and ε:

$$\begin{aligned}
&\partial_t\rho k+\nabla\cdot(\rho uk)-\nabla\cdot((\mu+\frac{\mu_T}{\sigma_k})\nabla k)-R:\nabla u+\rho\varepsilon\\
&=\underline{\frac{c_k\rho}{p}(u\cdot Ru)\nabla\cdot u},\\
&\partial_t\rho\varepsilon+\nabla\cdot(\rho u\varepsilon)-\nabla\cdot((\mu+\frac{\mu_T}{\sigma_\varepsilon})\nabla\varepsilon)-\frac{\varepsilon}{k}(c_{1\varepsilon}R:\nabla u-c_{2\varepsilon}\rho\varepsilon)\\
&\quad=\underline{c_{3\varepsilon}c_k\frac{\rho}{p}\frac{\varepsilon}{k}(u\cdot Ru)\nabla\cdot u}.
\end{aligned}$$

The constants take the following values:

$$c_\mu = 0.09 \quad \sigma_k = 1 \quad \sigma_\epsilon = 1.3$$
$$c_k = 1 \quad c_{1\epsilon} = 1.45 \quad c_{2\epsilon} = 1.92 \quad c_{3\epsilon} = 2$$

Initial conditions are given on $\rho, u, T, k, \varepsilon$ and boundary conditions should be given to u, T, k, ε on the whole boundary and on ρ on the inflow boundaries. Wall laws, low Reynolds number modifications or matching with boundary layer models are necessary near walls.

Remark

- In the definition of E, k may be neglected because usually $k << u^2/2$.
- In the definition of R, ρk may be set to zero because it is much smaller than p.
- In the equations for k and ε the terms containing $u.Ru$ are not frame invariant so most simulations are done without them (see Grasso[1989]).

APPENDIX A8: ALGEBRAIC AND ONE EQUATION MODELS

We present three models for turbulence in boundary layers with or without a small separation bubble. These may be used in conjunction with the $k - \varepsilon$ model which is not valid near the walls.
Matching these models with k and ε is done via μ_T with the two relations

$$\mu_T = c_\mu \rho \frac{k^2}{\varepsilon}, \quad \epsilon = \frac{k^{3/2}}{l_\epsilon}$$

where c_μ and l_ϵ are defined in §4.4.

1 The Cebeci-Smith model with a Goldberg correction
To take into account a possible separation Goldberg[1986], Cebeci et al [1974], Rogalo [1984], propose an alternative to the one equation model presented above (§4.5) for low Reynolds number or boundary layer regions.

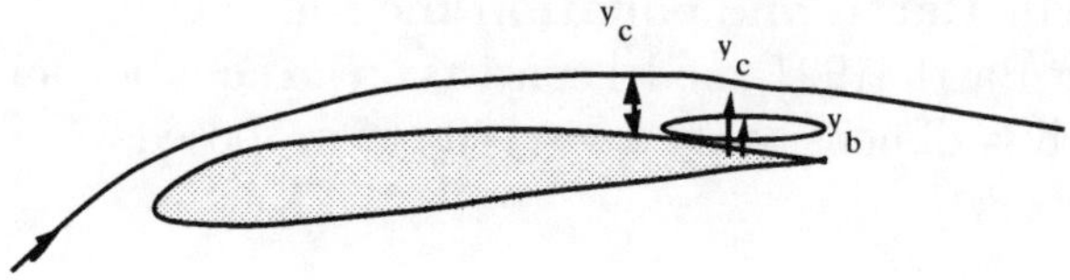

Figure 8.1

In this model, the eddy viscosity μ_T like in the Cebeci-Smith model, is a function of the distance to the wall (y) or to the backflow bubble edge $(y - y_b)$, in case of boundary layer separation. In the separation bubble, $(y \leq y_b)$, Goldberg's model is used:

$$\mu_T = 0.353 \rho u_s y_b \sqrt{\frac{\rho_w}{\rho}} (A \frac{y}{y_b} + B)(\frac{1 - \exp(-0.5(y/y_b)^2)}{1 - \exp(-0.5)}),$$

where ρ_w denotes the density at the wall and

$$c_\mu^* = 0.7, \quad A = -(\frac{c_\mu^*}{2})^{9/5}, \quad B = (\frac{c_\mu^*}{2})^{3/5} - A.$$

Outside the separation bubble but within the boundary layer ($y_b \leq y \leq y_c$), μ_T is given by

$$\mu_T = \rho l^2 |\omega|$$

with $\omega = \nabla \times u$ and above the boundary layer $(y \geq y_c)$, by

$$\mu_T = 0.0168 \rho u_e \delta_i (1 + 5.5(0.45 \frac{y - y_b}{y_{av}})^6)^{-1}$$

where

$$u_s = \sqrt{\frac{\mu_T}{\rho} \partial_y u_t |_{y=y_0}},$$

$$l = 0.41(y - y_b)(1 - exp(-\frac{y - y_b}{A})),$$

$$u_e \delta_i = \int_{y_b}^{y^*} y |\omega| dy \quad y_{av} = \frac{\int_{y_b}^{y^*} y^2 |\omega| dy}{\int_{y_b}^{y^*} y |\omega| dy},$$

with y_0 the value of y where $\partial_y u_t$ is maximum. Finally, y^* is taken such that $F(y^*) \approx 0.5 F_{\max}$ with $F(y) = y|\omega|(1 - \exp(-\frac{y}{A}))$.
Of course, for attached flows, we have only two levels in the previous scheme $(y_b = 0)$.

The Baldwin-Barth one equation model
The Baldwin-Barth [1991] model consists on an equation for a new quantity, r_ν which is dimensionaly a viscosity is proposed:

$$\partial_t r_\nu + u \nabla r_\nu = (c_{\epsilon_2} f_2 - c_{\epsilon_1}) G_1 \sqrt{r_\nu} P + (\nu + \frac{\nu_T}{\sigma_R}) G_1 \Delta r_\nu - \frac{1}{\sigma_\epsilon} \nabla \nu_T . \nabla r_\nu$$

The guiding principle is to obtain from the $k-\epsilon$ model (with low Reynolds number corrections) an equation for $r_\nu = k^2/\epsilon$ by combining the equations for k and for ϵ. The equation thus found has an additional term which drops when $P=\epsilon$.

As in the $k-\epsilon$ model, the production term is proportional to

$$P = \frac{\nu_T}{2}|\nabla u + \nabla u^T|^2 - \frac{2}{3}\nu_T|\nabla . u|^2.$$

with $\chi = 0.41$, $c_{\epsilon_1} = 1.2$, $c_{\epsilon_2} = 2.0$, $c_\mu = 0.09$,

$$\frac{1}{\sigma_\epsilon} = (c_{\epsilon_2} - c_{\epsilon_1})\frac{\sqrt{c_\mu}}{\chi^2}, \quad \sigma_R = \frac{\sigma_\epsilon}{4}.$$

While μ is the molecular diffusion and $\nu = \mu/\rho$, the turbulent diffusion is

$$\mu_T = c_\mu r_\nu D_1 D_2, \quad \nu_T = \frac{\mu_T}{\rho}$$

where D_1 and D_2 are two functions decaying exponentially away from the walls

$$D_1 = 1 - e^{-\frac{y^+}{A_1^+}}, \quad D_2 = 1 - e^{-\frac{y^+}{A_2^+}}$$

Suggested values are $A_1^+ = 26$, $A_2^+ = 10$. Then the low Reynolds number corrections are

$$f_2 = \frac{c_{\epsilon_1}}{c_{\epsilon_2}} + (1 - \frac{c_{\epsilon_1}}{c_{\epsilon_2}})(\frac{1}{\chi y^+} + D_1 D_2)[\sqrt{D_1 D_2} + \frac{y^+}{\sqrt{D_1 D_2}}(\frac{D_2}{A_1^+}e^{-\frac{y^+}{A_1^+}} + \frac{D_1}{A_2^+}e^{-\frac{y^+}{A_2^+}})].$$

It is suggested to take

$$G_1 = \frac{1 + F_R^4 + \frac{\epsilon_0}{B_2}}{\omega^4 + B_2 F_R^4 + \epsilon_0}$$

$$F_R = \frac{|\nabla r_\nu|^2}{r_\nu} - \frac{\chi^2}{c_\mu}|\omega|$$

where $\omega = \nabla \times u$ and $B_1 = 0.4$, $B_2 = 0.01$, $\epsilon_0 = 10^{-10}$.

At the wall one may take $r_\nu = 0$. Navier-Stokes equations are used with μ_T instead of μ.

Spalart-Allmaras' one equation model

The Spalart-Allamars' [1992] model is also an attempt to find an autonomous equation for the turbulent eddy diffusivity ν_T. But while Baldwin-Barth start from the $k - \epsilon$ model and make some assumption on k^2/ϵ, this model starts from simple flows such as shear flows and build an equation empirically from the usual production and dissipation terms. For simplicity we assume that $U|_\Gamma = 0$ at the wall. Then the turbulent diffusion is given by

$$\nu_T = \tilde{\nu} f_{v1}, \qquad \text{with } f_{v1} = \frac{\chi'^3}{\chi'^3 + c_{v1}^3}, c_{v1} = 7.1, \chi' = \frac{\tilde{\nu}}{\nu}$$

At the walls and at infinity $\tilde{\nu} = 0$, and in the computational domain, $\tilde{\nu}$ is given by

$$\partial_t \tilde{\nu} + u \nabla \tilde{\nu} - c_{b1}(1 - f_{t2}) \tilde{S} \tilde{\nu} - \frac{1}{\sigma}[\nabla.[(\nu + \tilde{\nu}) \nabla \tilde{\nu}] + c_{b2} |\nabla \tilde{\nu}|^2]$$

$$+[c_{w1} f_w - \frac{c_{b1}}{\chi^2} f_{t2}](\frac{\tilde{\nu}}{d})^2 - f_{t1} U^2 = 0$$

with

$$\tilde{S} = \frac{|\nabla u + \nabla u^T|}{2} + \tilde{\nu} \chi^{-2} d^{-2} f_{v2}$$

and the constants $\chi = 0.41$, $c_{b1} = 0.135$, $c_{b2} = 0.622$, $\sigma = 2/3$, $c_{t1} = 1$, $c_{t2} = 2$, $c_{t3} = 1.1$, $c_{t4} = 2$, $c_{w1} = c_{b1}/\chi^2 + (1 + c_{b2})/\sigma$, $c_{w2} = 0.3$, $c_{w3} = 2$. d= distance to the wall

$$f_w = g[\frac{1 + c_{w3}^6}{g^6 + c_{w3}^6}]^{\frac{1}{6}}, g = r + c_{w2}(r^6 - r), r = \frac{\nu_T}{\tilde{S} \chi^2 d^2},.$$

$$f_{v2} = 1 - \frac{\chi'}{1 + \chi' f_{v1}}$$

$$f_{t2} = c_{t3} \exp(-c_{t4} \chi'^2)$$

$$f_{t1} = c_{t1} g_t \exp(-c_{t2} \frac{\omega_t^2}{U^2}[d^2 + g_t^2 d_t^2]) \quad \text{with } g_t = min(0.1, \frac{U}{\omega_t \delta x})$$

where δx is the grid size along the wall.

CHAPTER 9

NUMERICAL TOOLS FOR COMPRESSIBLE K-EPSILON

1. INTRODUCTION

In this chapter we will describe a few numerical techniques for solving the $k-\varepsilon$ equations with particular emphasis on positivity and stability. We suppose that the mean flow variables (ρ, u, T) are known. For example, they are computed by one of the algorithm presented in Chapter 7.

In the following, in order to simplify the writing, the average quantities $\bar{\rho}, \tilde{u}, \tilde{T}$ will be denoted without bar or tilde.

We consider the frame invariant form of the $k-\varepsilon$ model where the molecular viscosity is neglected ($\mu = 0$):

$$\partial_t k + u\nabla k - \frac{c_\mu}{\rho}\nabla.(\rho\frac{k^2}{\varepsilon}\nabla k) = S_k,$$

$$\partial_t \varepsilon + u\nabla \varepsilon - \frac{c_\varepsilon}{\rho}\nabla.(\rho\frac{k^2}{\varepsilon}\nabla \varepsilon) = S_\varepsilon,$$

in $\Omega\times]0, T'[$, with the right hand side S_k and S_ε representing the production and dissipation of k and ε:

$$S_k = \frac{c_\mu}{2}\frac{k^2}{\varepsilon}|\nabla u + \nabla u^T|^2 - \varepsilon - \frac{2}{3}k\nabla.u - \frac{2}{3}c_\mu\frac{k^2}{\epsilon}|\nabla.u|^2,$$

$$S_\varepsilon = \frac{c_1}{2}k|\nabla u + \nabla u^T|^2 - c_2\frac{\varepsilon^2}{k} - \frac{2}{3}\frac{c_1}{c_\mu}\varepsilon\nabla.u - \frac{2c_1}{3}k|\nabla.u|^2.$$

Recall that the constants have the following values:

$$c_\mu = 0.09,\ c_\epsilon = 0.07,\ c_1 = 0.128,\ c_2 = 1.92$$

It is convenient to introduce the following notations:

$$\begin{aligned} D_t &= \partial_t + u\nabla, \\ D &= \nabla \cdot u, \\ F &= \frac{1}{2}[\nabla u + \nabla u^T|^2 - \frac{2}{3}D^2, \end{aligned},$$

and rewrite the system as

$$D_t k - \frac{c_\mu}{\rho}\nabla \cdot (\rho\frac{k^2}{\varepsilon}\nabla k) = S_k = c_\mu \frac{k^2}{\varepsilon}F - \frac{2}{3}kD - \varepsilon$$

$$D_t \varepsilon - \frac{c_\varepsilon}{\rho}\nabla \cdot (\rho\frac{k^2}{\varepsilon}\nabla \varepsilon) = S_\varepsilon = c_1 kF - \frac{2c_1}{3c_\mu}\varepsilon D - c_2\frac{\varepsilon^2}{k}.$$

2. POSITIVITY OF K AND EPSILON

For physical reasons, it is important that the model guarantees the positivity of k and ε if they are positive initially. We will give here a positivity argument similar to the one given in chapter 6. We suppose that the system has a smooth solution for given positive data and we introduce a new variable $\theta = \frac{k}{\varepsilon}$ (the turbulent time scale), the θ equation can be obtained from those of k and ε, as follows:

$$D_t\theta = \frac{\partial \theta}{\partial t} + u \cdot \nabla\theta = \frac{1}{\varepsilon}D_t k - \frac{k}{\varepsilon^2}D_t\varepsilon = -a\theta^2 F + b\theta D + c + D_{iff},$$

where

$$a = (c_1 - c_\mu), \quad b = \frac{2}{3}(\frac{c_1}{c_\mu} - 1), \quad c = c_2 - 1,$$

with D_{iff} being the contribution from the viscous terms in the k and ε equations :

$$D_{iff} = (c_\mu + 2c_\epsilon)\theta\nabla k\nabla\theta + c_\varepsilon\frac{k\theta}{\rho}\nabla\theta\nabla\rho + c_\varepsilon k|\nabla\theta|^2 + c_\varepsilon k\theta\Delta\theta$$

$$+ (c_\mu - c_\varepsilon)(\theta^2\Delta k + \text{sign}(k)\theta^2|\nabla\sqrt{|k|}|^2 + \frac{\theta^2}{\rho}\nabla\rho\nabla k)$$

Now, through the θ equation, we can prove the following

Theorem
Assume k and ε smooth at all times and positive at time zero. Then they stay positive at all times.

Proof: Because $c_\mu \geq c_1$ and $c_2 \leq 1$, θ will stay positive and bounded when there are no diffusion terms. The dynamical part of the θ equation is

$$D_t\theta = -aF\theta^2 + bD\theta + c,$$

with

$$\theta(0) = \frac{k(0)}{\varepsilon(0)} = \theta_0.$$

This is a Riccati equation. If we suppose F and D constant along characteristic curves then

$$\theta(t) = \frac{\theta_1 - K\theta_2 e^{-st}}{1 - Ke^{-st}}$$

where $s = (b^2D^2 + 4acF)^{\frac{1}{2}} \geq 0$, $K = (\theta_0 - \theta_1)/(\theta_0 - \theta_2)$ and θ_1 and θ_2 are roots of $-aF\theta^2 + bD\theta + c$. The non constant case being similar, if θ is initially positive, it stays positive and bounded. This argument is extended without difficulty to the non constant F, D case.
Also, θ cannot become negative even in the presence of the viscous terms D_{iff}. Call t_m the first time when there exists x_m such that $\theta(x_m, t_m) = 0$; because we have positive boundary and initial conditions for θ, x_m is strictly inside Ω and so at this minimum we must have $\nabla\theta = 0$ and $\Delta\theta \geq 0$.
Notice that every term in D_{iff} is multiplied by θ or $\nabla\theta$. So $D_{iff}(x_m, t_m) = 0$. Hence $\frac{d\theta}{dt} = c > 0$ and θ will increase again beyond t_m.

Furthermore, introducing θ in the k equation, we have

$$\frac{\partial k}{\partial t} + u \cdot \nabla k - \frac{1}{\rho}\nabla \cdot ((c_\mu \rho k\theta)\nabla k) = c_\mu k\theta F - \frac{2}{3}kD - \frac{k}{\theta}.$$

Denote again by (x_m, t_m) the first point at which k is minimum. Suppose this minimum is 0. When positive Dirichlet boundary condition are added to these equations, x_m cannot be on the boundary. So at

(x_m, t_m) we will have $k = 0, \nabla k = 0, \Delta k \geq 0$. But, we have shown that θ is always strictly positive so the equation for k reduces to

$$\partial_t k \geq 0.$$

Thus k is never negative.
Exactly the same reasoning can be made with ε because its equation is, in terms of θ,

$$\partial_t \varepsilon + u \cdot \nabla \varepsilon - \frac{1}{\rho} \nabla \cdot (c_\varepsilon \rho k \theta \nabla \varepsilon) = c_1 \varepsilon \theta F - \frac{2c_1}{3c_\mu} \varepsilon D - c_2 \frac{\varepsilon}{\theta}.$$

3. STABILITY OF THE K-EPSILON SYSTEM

Another major difficulty, if we have to use the $(k - \varepsilon)$ model, is that there is the possibility of an exponential growth for k and ε. Indeed, if $t \to \infty$ and the Dirichlet conditions on θ are constant and equal to θ_1 then , $\theta \to \theta_1$ and the k and ε equations tend to

$$D_t k = k(c_\mu \theta_1 F - \frac{2}{3} D - \frac{1}{\theta_1}),$$

$$D_t \varepsilon = \varepsilon(c_1 \theta_1 F - \frac{2c_1}{3c_\mu} D - \frac{c_2}{\theta_1}).$$

Therefore, if F is sufficiently large, both k and ε grow exponentially.

As shown, $\theta = \frac{k}{\varepsilon}$ has a good numerical behaviour (bounded and positive), the idea is then to find another combination φ of k and ε, which is dynamically stable ($D_t \varphi \leq 0$). Because of the similar behaviour of k and ε, we look at the dynamical part of their equations and try to find $\varphi = k^\alpha \varepsilon^\beta$ with α and β real.

The φ equation comes from those of k and ε as for θ. Without the viscous terms it is

$$D_t \varphi = k^\alpha \varepsilon^\beta (aF\theta + bD + \frac{c}{\theta}),$$

where $a = \alpha c_\mu + \beta c_1$, $b = -\frac{2}{3}(\alpha + c_1 \beta / c_\mu)$ and $c = -\alpha - \beta c_2$.

Thus φ will decrease ($D_t\varphi \leq 0$) if a and c are negative and if b is zero (the sign of D is not known). Now, $b = 0$ implies $\alpha = -c_1\beta/c_\mu$ and gives

$$D_t\varphi = -\beta(c_2 - \frac{c_1}{c_\mu})\frac{\varphi}{\theta}.$$

Recall that $\frac{c_1}{c_\mu} = 1.44 < c_2$; so any positive value for β will work but the closest choice to the incompressible values ($\varphi = \varepsilon^2/k^3$ in §IV.4.2) is $\beta = 2.1$, $\alpha = 3$ or $\beta = 2$ and $\alpha = 2.88$, giving, with the second choice:

$$\varphi = \frac{\varepsilon^2}{k^{2\frac{c_1}{c_\mu}}} \approx \frac{\varepsilon^2}{k^{2.88}}.$$

Remark
We can make a similar analysis for the low Reynolds number versions of the k, ε model; in these models, the constants (c_μ, c_1, c_2) are multiplied by positive functions (f_μ, f_1, f_2) which depend on $R_y = \rho k^2/\mu\varepsilon$ and $R_t = \rho\sqrt{k}y/\mu$ two local Reynolds numbers (Patel et al [1989]). Hence, in the equation for φ, we have now $a = \alpha c_\mu f_\mu + \beta c_1 f_1, b = -\alpha - c_1 f_1 \beta/c_\mu$ and $c = -\alpha - \beta f_2 c_2$. Because $0 < f_\mu \leq 1$, β must be negative and if we take $\alpha = f_1 c_1/c_\mu$ we have

$$D_t\varphi = (c_2 f_2 - \frac{c_1 f_1}{c_\mu})\frac{\varphi}{\theta}.$$

So for φ to decrease, we must have $f_2 \leq \frac{c_1 f_1}{c_2 f_2}$; but this is not always true because $1 \leq f_1$ and $0 < f_2 \leq 1$.

4. NUMERICAL ALGORITHMS

In this section also, we assume ρ, u and T known, computed for example by one of the techniques presented in Chapter 7.

4.1. Finite Element/Finite Volume method

To compute k and ε, we use the multistep algorithm described earlier where convection is performed first and then diffusion.

a) The transport step

Without the diffusion terms, the $k - \varepsilon$ equations are

$$D_t k = S_k = -c_\mu \frac{k^2}{\varepsilon} F - \varepsilon - \frac{2}{3} k D,$$

$$D_t \varepsilon = S_\varepsilon = c_1 k F - c_2 \frac{\varepsilon^2}{k} - \frac{2c_1}{3c_\mu} \varepsilon D.$$

We multiply the equations by ρ and use an identity due to the equation of ρ:

$$\rho D_t k = \partial_t(\rho k) + \nabla \cdot (\rho u k).$$

Then we discretise this term explicity in time, after integration over the cells σ (see Chapter 7)

$$\int_\sigma \rho D_t k \approx \frac{1}{\delta t} \int_\sigma \rho^n (k^{n+1} - k^n) + \int_{\partial\sigma} (\rho^n u^n \cdot \vec{n} k^n)_d.$$

The subscript d stands for upwinding; the value chosen is the one on the element upstream with respect to u.

Then we decompose the source terms in the following way

$$S_k = S_k^+ - S_k^-, \quad S_k^+ = c_\mu \frac{k^2}{\varepsilon} F + \frac{2}{3} D^-, \quad S_k^- = \varepsilon + \frac{2}{3} k D^+,$$

$$\tilde{S}_k^- = \frac{S_k^-}{k},$$

where $D^\pm$ are the positive and negative parts of D. Then, S_k^+ and $\tilde{S}_k^-$ are both positive. Both of them are treated explicitly, the values for k and ε being those at the beginning of the time step, k^n, ε^n. We find at t^{n+1} values for k and ε, which we denote by $k^{n+\frac{1}{2}}$ and $\varepsilon^{n+\frac{1}{2}}$ given by

$$\int_\sigma \rho^n k^{n+1/2} (1 + \delta t \tilde{S}_k^{n^-})$$
$$= \int_\sigma \rho^n (k^n + \delta t S_k^{n^+}) - \delta t \int_{\partial\sigma} (\rho^n u^n \cdot \vec{n} k^n)_d$$

and similarly for ε.

This treatment of the source terms (explicit for the positive part and semi-implicit for the negative part) tempers the exponential growth of k and ε when F is large and slows down the decay of the same quantities when F is small.

b. The diffusion step

To obtain $k^{n+1}, \varepsilon^{n+1}$, we solve with an implicit scheme the following

$$\rho\partial_t k = c_\mu \nabla \cdot [\rho(\frac{k^2}{\varepsilon})\nabla k],$$

$$\rho\partial_t \varepsilon = c_\varepsilon \nabla \cdot [\rho(\frac{k^2}{\varepsilon})\nabla \varepsilon],$$

on a time interval $[t^n, t^{n+1}]$ starting from the value obtained at the end of the transport step $k^{n+\frac{1}{2}}, \varepsilon^{n+\frac{1}{2}}$. The viscosities and the density are evaluated explicitly, i.e. at time t^n.

$$\rho^n \frac{1}{\delta t}(k^{n+1} - k^{n+1/2}) - c_\mu \nabla \cdot [\rho(\frac{{k^n}^2}{\varepsilon^n})\nabla k^{n+1}] = 0.$$

Discretization in space is done with a $P1$ continuous discretization for k and ε on triangles and piecewise constant for the viscosities and the density. The viscosities are piecewise constant over each triangles but the densities are piecewise constant over the cells around each vertex so as to use mass lumping. Positivity of the overall algorithm is then obtained if there is no obtuse angle in the mesh Ikeda[1976], Ciarlet[1978].
Since this step requires the solution of a linear system we may as well mix both step in order to improve the stability of the convection step. At the cost of making the linear system non-symmetric we can treat the convection term implicity (while keeping upwinding). Then, and only then, it is possible to show unconditional stability and positivity.

4.2 Characteristic-Galerkin for the Convection Step

a. Transport of $k - \varepsilon$

Positivity is more robust when the Characteristic method (Benque et al[1982], Pironneau[1989], Douglas et al[1982]) is used to discretize the total derivative

$$D_t k(x,t^{n+1}) \approx \frac{1}{\delta t}[k(x,t^{n+1}) - k(X^n(x),t^n)],$$

and we retain the idea of Benqué et al [1980] to integrate the source terms on the characteristics because for a fixed time step the source terms have large variations on the travel path $]x, X^n(x)[$.

So we consider the characteristic issued from x, i.e. the solution of the ODE below with boundary conditions at the end of the interval:

$$\frac{d}{d\tau}X(\tau) = u(\tau, X(\tau)), \quad X(t^{n+1}) = x, \quad \tau \in [t^n, t^{n+1}],$$

and call $X^n(x) = X(t^n)$.
With the same splitting for the source terms we obtain

$$[1 + \int_{t^n}^{t^{n+1}} \tilde{S}_k^{n^-}(X(\tau))d\tau]k^{n+\frac{1}{2}} = k^n o X^n + \int_{t^n}^{t^{n+1}} S_k^{n^+}(X(\tau))d\tau$$

$$[1 + \int_{t^n}^{t^{n+1}} \tilde{S}_\varepsilon^{n^-}(X(\tau))d\tau]\varepsilon^{n+\frac{1}{2}} = \varepsilon^n o X^n + \int_{t^n}^{t^{n+1}} S_\varepsilon^{n^+}(X(\tau))d\tau.$$

where

$$S_\varepsilon^+ = c_1 kF + \frac{2c_1}{3c_\mu}\varepsilon D^-, \quad \tilde{S}_\varepsilon^- = c_2\frac{\varepsilon}{k} + \frac{2c_1}{3c_\mu}D^+$$

Notice that the characteristics are only computed once for each quadrature points and that we integrate four different quantities $(S_k^+, \tilde{S}_k^-, S_\varepsilon^+, \tilde{S}_\varepsilon^-)$ along these characteristics.

b. Transport of $\theta - \varphi$
The previous approach guarantees also the positivity of k and ε, but it does not prevent the possibility of a blow up for k and ε. To improve the robustness and stability of the method, we introduce θ and φ as presented previously. Thus, knowing k^n and ε^n, we evaluate $\varphi^n = [(k^n)^{-\frac{c_1}{c_\mu}}\varepsilon^n]^\beta$ and $\theta^n = \frac{k^n}{\varepsilon^n}$. Now, using the integration along the characteristics as above we perform the convection step on θ, φ rather than on k, ε:

$$[1+\int_{t^n}^{t^{n+1}} \tilde{S_\varphi^-}(X(\tau))d\tau]\varphi^{n+\frac{1}{2}} = \varphi^n o X^n,$$

and

$$[1+\int_{t^n}^{t^{n+1}} \tilde{S_\theta^-}(X(\tau))d\tau]\theta^{n+\frac{1}{2}}(x) = \theta^n o X^n + \int_{t^n}^{t^{n+1}} S_\theta^+(X(\tau))d\tau.$$

with

$$\tilde{S}_\varphi^- = \frac{\beta}{\theta}(c_2 - \frac{c_1}{c_\mu}),$$

$$\tilde{S}_\theta^- = \frac{2}{3}(\frac{c_1}{c_\mu} - 1)D^- + (c_1 - c_\mu)F\theta, \quad S_\theta^+ = c_2 - 1 + \frac{2}{3}(\frac{c_1}{c_\mu} - 1)D^+\theta$$

Now $k^{n+\frac{1}{2}}$ and $\varepsilon^{n+\frac{1}{2}}$ are recovered from the definition of θ and φ:

$$k^{n+\frac{1}{2}} = (\theta^{n+\frac{1}{2}})^{\frac{1}{1-\frac{c_1}{c_\mu}}}(\varphi^{n+\frac{1}{2}})^{\frac{1}{\beta(1-\frac{c_1}{c_\mu})}}$$

and

$$\varepsilon^{n+\frac{1}{2}} = (\theta^{n+\frac{1}{2}})^{\frac{1}{\frac{c_\mu}{c_1}-1}}(\varphi^{n+\frac{1}{2}})^{\frac{1}{\beta(1-\frac{c_1}{c_\mu})}}.$$

It has been shown above that φ and θ stay bounded and positive, so the algorithm will be robust.

Remark
An even better method, though more expensive, is to assume all coefficients in the equations constants during the time step and solve both equations analytically. This gives

$$\theta^{n+1} = \frac{\theta_1 - K\theta_2 e^{-s\delta t}}{1 - Ke^{-s\delta t}}$$

where $s, theta_1, theta_2$ are as in §2 and $K = (\theta^n - theta_1)/(\theta^n - \theta_2)$. Similarly

$$\varphi^{n+1} = \varphi^n \exp(-\beta(c_2 - \frac{c_1}{c_\mu})\frac{\delta t}{\theta^n})$$

4.3 A Least Square SUPG Method

Alternatively SUPG can be applied to the system made of Navier-Stokes equations plus $k-\varepsilon$ equations. Thus it is a generalization of the technique presented in Chapter 7 for the Navier-Stokes equations.
Recall that Navier-Stokes and $k-\varepsilon$ equations can be written as:

$$\partial_t W + \nabla \cdot F(W) - \nabla \cdot G(W, \nabla W) = S(W),$$

where, in two dimensions, $W = (\rho, \rho u_1, \rho u_2, \rho E, \rho k, \rho\varepsilon)$, F and G are the convective and viscous terms and S contains the zero order terms in the k and ε equations.

One possibility is to apply SUPG to the full system. Another, equally efficient, is to apply SUPG to the Navier-Stokes equation and to the $k-\varepsilon$ equations separately. We present the second method.
Since SUPG for compressible Navier-Stokes equations has been presented in §7.3, we need only to show how SUPG is applied to system above when $W = (\rho k, \rho\varepsilon)$ and

$$F(W) = (\rho u k,\ \rho u \varepsilon),$$

$$G(W, \nabla W) = (c_\mu \rho \frac{k^2}{\varepsilon} \nabla k,\ c_\varepsilon \rho \frac{k^2}{\varepsilon} \nabla \varepsilon),$$

and

$$S(W) = (\mu_t F - \frac{2}{3} \rho k D - \rho\varepsilon,\ c_1 F \rho k - \frac{2}{3} \frac{c_1}{c_\mu} \rho \varepsilon D - c_2 \rho \frac{\varepsilon^2}{k})^T.$$

The same equation in quasi-linear form is

$$\partial_t W + A_i \partial_i W - \partial_i (K_{ij} \partial_j W) = S(W),$$

where

$$A_i = \frac{\partial F_i}{\partial W} \quad \text{and} \quad K_{ij} = (\frac{\partial W}{\partial x_j})^{-1} N_i.$$

Let $\{T_i\}_1^{n_T}$ be a triangulation of Ω; let $\Omega = \cup_i T_i$. Let H_h be a space of continuous piecewise polynomial functions on the triangulation. In case of Dirichlet conditions on the whole of Γ, the finite element interpolation space V_h and the test function space V_{0h} are defined by:

$$V_{0h} = \{W_h \in [H_h]^2 : \quad W_h|_\Gamma = 0\},$$

and

$$V_h = \{W_h \in [H_h]^2 : \quad W_h|_\Gamma = g_h\},$$

where g_h contains the Dirichlet boundary conditions.
The SUPG finite element formulation can be written as follows: find $W_h \in V_h$ such that for all $\Phi_h \in V_{0h}$

$$\int_\Omega [\Phi_h(\frac{W_h^{n+1} - W_h^n}{\delta t} + A_{h_i}^n \partial_i W_h^{n+1} + \tilde{S}_h^{n^-} W_h^{n+1}) + \partial_i \Phi_h K_{h_{ij}}^n \partial_j W_h^{n+1}]$$
$$+ \int_\Omega \tau A_{h_k}^{n^T} \partial_k \Phi_h [\frac{1}{\delta t}(W_h^{n+1} - W_h^n) + A_{h_i}^n \partial_i W_h^{n+1} - \partial_i (K_{h_{ij}}^n \partial_j W_h^{n+1})]$$
$$+ \int_\Omega \delta \; \partial_i \Phi_h \partial_i W_h^{n+1} = \int_\Omega [\Phi_h + \tau A_{h_k}^{n^T} \partial_k \Phi_h] S_h^+(W_h^n).$$

Here the two first integrals represent the Galerkin part of the formulation, the third integral the SUPG term and the fourth the discontinuity capturing term. Some integrals involve derivatives of discontinuous functions on the edges of the triangulation; these are understood as sums of integrals over the triangles.

The contribution of the right hand side S has been distributed on the left hand side and on the right hand side of the equation. Moreover, S is discretized semi-implicity in time. Indeed S is rewritten as

$$S(W_h) \approx S_h^+(W_h^n) - W_h^{n+1} \tilde{S}_h^-(W_h^n),$$

where S^+ and $\tilde{S}^-$ are diagonal matrices defined by:

$$S_h^{\pm} = \mathrm{diag}[S_i^{\pm}], \quad \tilde{S}^- = \mathrm{diag}[\frac{S_i^-}{W_i}].$$

As before f^+ and f^- are the positive and negative part of f. As we have seen in chapter 5, this splitting of the right hand side in positive and negative parts improves the stability and the positivity of the scheme.

The SUPG parameter, τ, and shock-capturing parameter, δ, are defined as in Mohammadi [1993].

$$\tau_k = 0.5 \; \max_i (\frac{h_i |\beta_{ki}|}{c + |u.\beta_k|}), \quad \tau_\varepsilon = 0.5 \; \max_i (\frac{h_i |\beta_{\varepsilon i}|}{c + |u.\beta_\varepsilon|}),$$

where h_i the element length in the ith direction, and c is the local speed of sound . Moreover, the coefficients β_i are defined by:

$$\beta_k = \frac{\nabla(\rho k)}{||\nabla(\rho k)||_0} \quad \beta_\varepsilon = \frac{\nabla(\rho\varepsilon)}{||\nabla(\rho\varepsilon)||_0}.$$

The Shock-capturing parameter δ is a diagonal matrix defined by:

$$\delta = \text{diag}[\delta_k, \delta_\varepsilon],$$

where δ_k and δ_ε are defined by:

$$\delta_k = \frac{[\sum_{i,j}(A_{i_{1j}}\partial_i W_j{}^2]^{\frac{1}{2}}}{||\nabla(\rho k)||_0} \quad \text{and} \quad \delta_\varepsilon = \frac{[\sum_{i,j}(A_{i_{2j}}\partial_i W_j{}^2]^{\frac{1}{2}}}{||\nabla(\rho\varepsilon)||_0}.$$

As in chapter 7, the scheme can be improved by switching the SUPG and shock capturing terms in zones where they are not necessary (set $\delta = 0$ whenever the solution is smooth, then set $\tau = 0$ when the mesh Peclet number is small).

CHAPTER 10

NUMERICAL RESULTS AND EXTENTIONS

Summary
In this chapter, we give first some simple test cases which can be used to validate a $k-\varepsilon$ solver, then we give some numerical results obtained with our $k-\varepsilon$ solver developed using the finite volume / finite element and characteristic-Galerkin techniques presented in chapters 7 and 9.

1. FLOW OVER A FLAT PLATE

The problem presented is a turbulent flow over a flat plate at $M_\infty = 0.5$ and $Re = 3.2 \times 10^6$. The mesh used for this case contains 3570 nodes. The first point is placed at 10^{-5}m. The length of the plate is 1 meter and the height of the computational domain is 0.2 m. At the inlet boundary we have:

$$\rho = 1,\ u_1 = 1,\ u_2 = 0,\ T = \frac{1}{\gamma(\gamma-1)M_\infty^2},\ k = \varepsilon = 10^{-4}.$$

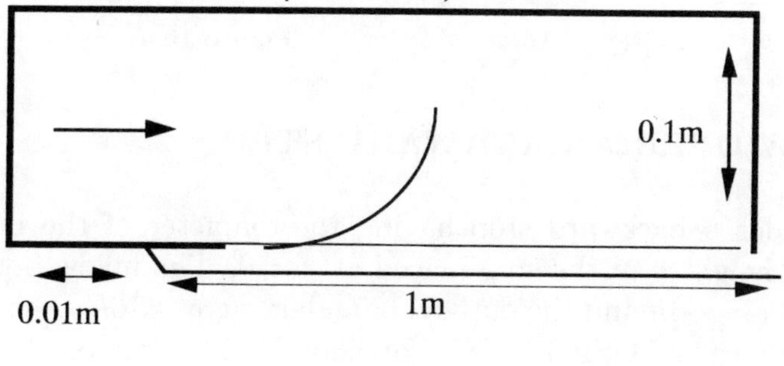

Figure 10.1

The boundary condition on the top and the outlet boundaries are those of an Euler calculation. In this case, the low-Reynolds regions has been computed by the two-layers approach presented in Chapter 8. Figures 10.2, 10.3 show the velocity profile and the skin friction coefficient. The computed horizontal velocity and skin friction coefficient are compared with the theoretical $\frac{1}{7}$ power law and $0.0375 Re\, x^{-1/6}$ respectively Cousteix [1990]. In both cases, the agreement with theoretical solutions is quite good. As we have computed the flow up to the wall, the power law profile does not match the computed velocity profile in the near-wall region. For this computation we assumed that the transition from laminar to turbulent occurs between $x = 0.12m$ and $x = 0.15m$. For this reason, we multiply the eddy viscosity by a function varying from zero to one through this zone. This explain the fast variation of the skin friction in this area.

2. FLOW IN A CAVITY

The domain is shown on figure 10.4 and the quadrilateral mesh (6000 triangles) on figure 10.5. The fluid flows from left to right and creates an eddy in the cavity. At time zero all variables are set to zero except $k = \varepsilon = 10^{-2}$ or 10^{-3} (the results are the same). The inflow boundary is the first vertical segment on the left; the boundary conditions are blunt (constant), the Reynolds number is 10^5. On the outflow boundary (the last vertical segment on the right) Neumann conditions are imposed. Near the walls a one equation model is used. Results are shown on figures 10.6,10.7.

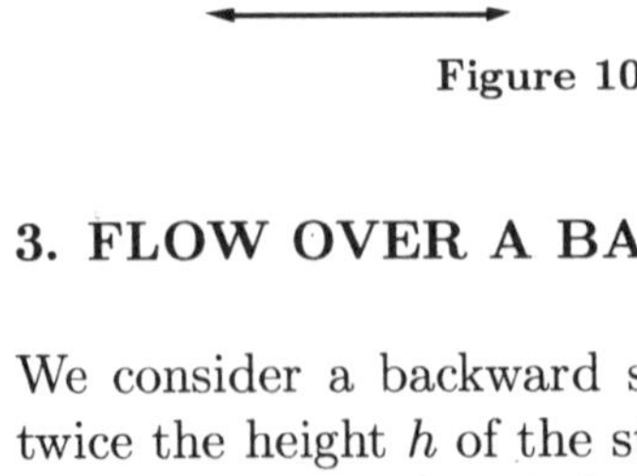

Figure 10.4

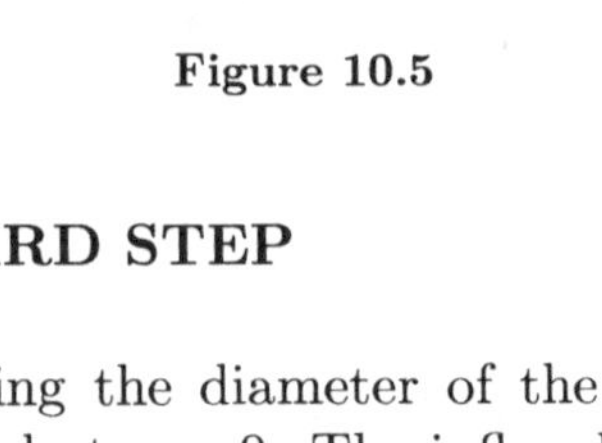

Figure 10.5

3. FLOW OVER A BACKWARD STEP

We consider a backward step having the diameter of the inlet canal twice the height h of the step placed at $x = 0$. The inflow boundary is placed at $x = -4h$ and the outflow boundary at $x = 16h$. This test case comes from the AFOSR-HTTM Workshop held at Stanford in 1980 and 1981 (AFOSR[1981]).

The Reynolds number is 44580 and the inflow Mach number 0.1. A constant profile is given for all the variables at the inlet boundary.

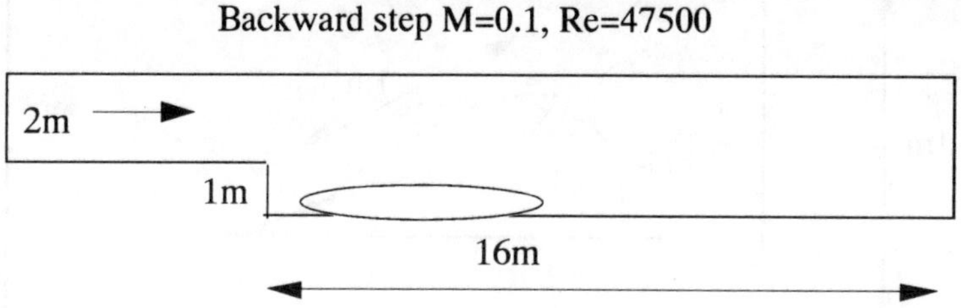

Figure 10.8

This test case is a classical test to prove the inability of the $k-\varepsilon$ model to predict the correct length of a recirculation bubble. For this case we have used the wall-law of Chapter 8. The wall-law under predicts the level of turbulent quantities in regions of separation. This is due to the fact that in these regions the friction at the wall vanishes. For the wall-law, δ is set to 0.05. Computed u, k, ε and μ_T along the upper and lower walls are shown (figures 10.9-10.12). As expected, the computed length of the recirculation bubble is slightly more than 5 against an experimental value of 7 and we observe very low levels for the turbulent quantities in the detachment and reatachment regions.

4. FLOW PAST AN AIRFOIL

We present a transonic configuration around two airfoils: a RAE2822 and a NACA 0012 airfoil

4.1 Flow around a NACA0012

The test is taken from from the Viscous Transonic Airfoil Workshop (AIAA[1987]) held at the 25th AIAA Reno Meeting.

Experimental data for the pressure coefficient on the airfoil surface are available (Harris[1981]). The free stream Mach and Reynolds numbers are $M_\infty = 0.55, Re_\infty = 9 \times 10^6$ and the angle of attack is 8.34 degrees.

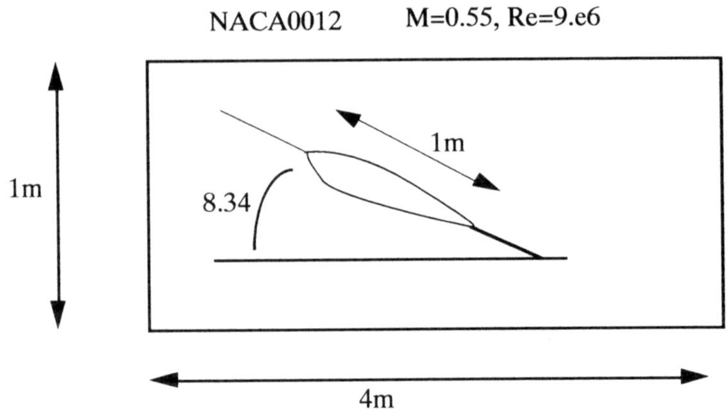

Figure 10.13

This is a complex configuration for the following reasons: a supersonic bubble is well forward on the airfoil upper surface, the flow is slightly separated at the foot of the shock and the angle of attack is about one degree below the maximum lift value. Iso Mach, μ_T and k contours are shown on figure 10.14, 10.15, 10.16. This case has been computed with a two-layer approach. The transition from laminar to turbulent has not been taken into account in the computation.

4.2 Flow around a RAE2822

For this profile experimental data are also available (cf Holst [1987]). The condition of the experiment are: $M_\infty = 0.75$, 2^0 of incidence, $Re = 6.2\ 10^6$; The computation was done with the Finite Volume / Finite Element approach and a wall law.
Figure 10.17 (see back cover of the book) shows the μ_T. Figure 10.18 shows the pressure on the wall, Figure 10.19 the friction coefficent $(\rho_w u_\tau^2/0.5\rho_\infty u_\infty^2)$ and Figure 10.20 the displacement thickness $(\int_0^\delta(1 - u/u_E)$ where δ is the boundary layer thickness and u_E the velocity at the edge of the boundary layer. Finally Figure 10.21 shows a velocity profile in a cross section perpandicular to the profile ($x/c = 90\%$, where c is the cord length). The first point is approximately at $y^+ = 100$.

5. HYPERSONIC COMPRESSION CORNER

We study a 15° hypersonic compression corner from the Hermes Workshop held in 1991 in Antibe. The free stream mach number is $M_\infty = 10$ and the Reynolds number per meter $Re = 9.10^6$.

The free stream temperature is $T_\infty = 50K$ and the wall temperature $T_w = 288K$. The mesh has 4587 nodes and 8832 triangles. The first spacing normal to the wall is fixed at 10^{-5}.

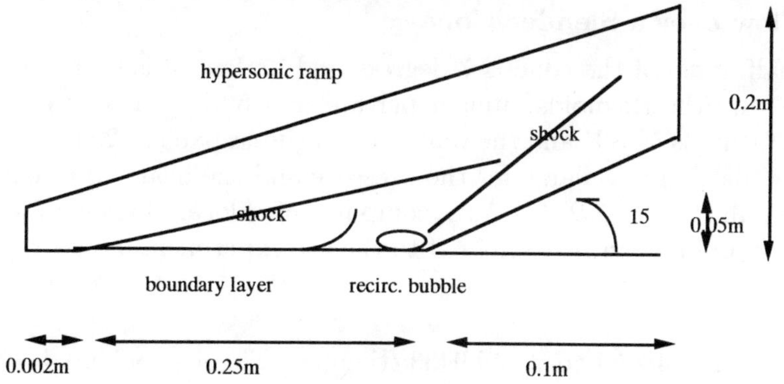

Figure 10.22

On this mesh we have done two computations. A completely laminar computation (i.e. without any turbulence modeling) and another computation using our $k - \varepsilon$ solver. In the second case, the turbulence model is started at the corner ($x = 0.25$). In the turbulent case we have a smaller recirculation zone and the levels of the pressure and the heat flux coefficients after the corner are higher than for the laminar computation. Plots of the pressure and heat flux coefficients are given in figures 10.23, 10.24. These results are compared with data from Delery[1990]. Better agreement is obtained with the turbulent computation (especially for the heat flux coefficient after the corner). These results are interesting because the nature of the flow is not well-known (laminar, transitional or turbulent).

6. AXISYMMETRIC FLOWS

We now describe some test cases of axisymmetric turbulent hypersonic flows computed with our $k - \varepsilon$ solver and two-layer techniques near the wall.
The equations are given in appendix. For this case we suppose $u_\theta = 0$. More details can be find in Mohammadi[1993].
We consider, the test cases 1 and 2 from the Antibe Workshops on Hypersonic Flows for Reentry Problems (1990 and 1991). The first case is the flow over a slender cone. and the second case consists of the base flow over an axisymmetric downward step behind the cone. In this case, the flow is fully detached, turbulent and hypersonic (problem 2). The

computations are compared to available experimental data (Denman et al. [1990]) and occasionally to other computations.

6.1 Flow over a Slender Cone

The half angle of the cone is 7 degrees and its length is 0.578 m. M_∞ is 9.16 and the Reynolds number per meter is 5.510^7. The free stream temperature is 59.8 K and the wall temperature is fixed at 290 K. Experimental data are available for the pressure and the heat flux coefficient at the wall ($C_h = -\kappa\partial_n T$). Also, computational local Mach number and velocity profiles can be compared to the the experimental ones through the boundary layer.

The mesh has 4865 nodes and 9433 triangles. The first point is $5.10^{-6}m$ away from the wall. We have done two computations. A first one by taking into account the transition to turbulence and another computation where the flow is supposed fully turbulent (this means that the turbulence model was activated everywhere). In the first case, the transition from laminar to turbulent is arbitrarily assumed to occur between $x = 0.1m$ and $x = 0.2m$ from the leading edge and this is taken into account, as for the flat plate case, by multiplying the eddy viscosity by a function varying from 0 to 1 through this zone. Experiment shows that the transition occurs between $x = 0.6m$ and $x = 0.1m$. We found that small modifications of this position has little effect on the turbulent flow.

The time-averaged static pressure distribution, normalized by the free stream static pressure $p_\infty = 1/\gamma M_\infty^2$, computed on the cone surface is compared to the experimental recorded values in picture 10.25.

Both computations give approximately the same result for the fully turbulent region. The pressure at the wall is then overestimated by about 10 percent. These results are in agreement with the results of the other contributors of the workshop. In order to find the cause of this discrepancy, one of the contributors (Lawrence[1990]) studied several kind of sensitivity, but no explanation could be found. He studied the following effects:
- Sensitivity to the variation of the inflow mach number gradient,
- Sensitivity to grid spacing,
- Sensitivity to wall temperature,
- Sensitivity to turbulence model.

Picture 10.26 shows the computed and experimental heat flux coefficients ($C_h = -\kappa \partial_n T.$). Because the transition is badly imposed, the maximum of the heat flux coefficient is displaced downstream. In the fully turbulent region, the heat flux is underestimated by both computations.

Computed velocity profile and Mach number across the boundary layer are compared to experimental data at $x = 0.18m$ upstream of the base (figures 10.27,10.28). The agreement is excellent in both cases.

For the base flow case, we begin the computation at $x = 0.18m$ upstream of the base and we take as inlet boundary condition the values obtained in the cone case.

6.2 Turbulent Base Flow

The previous cone is mounted on a shaft with diameter 0.04m smaller than the base of the cone. Therefore, the flow separates and a recirculation zone exists. The flow conditions at infinity and at the wall are the same than for the cone. Again, experimental data [6] for the pressure and the heat flux at the wall and the velocity and the Mach number profiles through the boundary layer are available. We have also compared our results to the contribution of M.P.Netterfield (Netterfield [1990]).

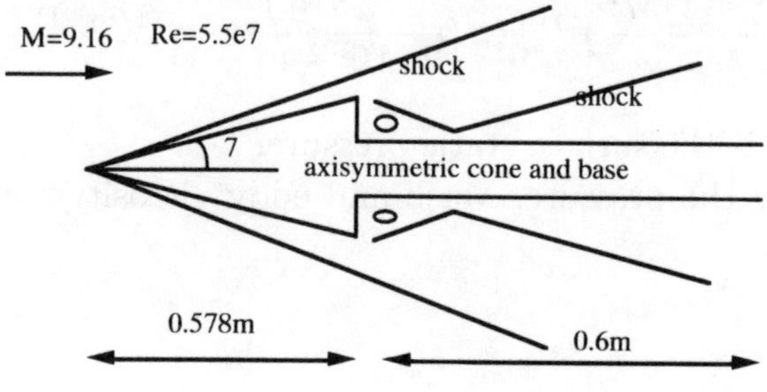

Figure 10.29

For this case, as for the 2D backward step, we expect the $k-\varepsilon$ model to give erroneous results in the reattachment region. The mesh has 9636 nodes and 18878 triangles. As inlet boundary conditions, we use the results obtained in the cone case. The outlet boundary is placed at $x = 1.2m$. Neumann boundary conditions are used at this boundary for all variables. At the top boundary free stream Dirichlet boundary conditions are imposed.

The time-averaged static pressure distribution, normalized by the free stream static pressure $p_\infty = 1/\gamma M_\infty^2$, computed in the base region is compared to the experimentally recorded values and to the results of Netterfield in picture 10.30. In the plot, $x = 0$ refers to the cone/sting corner. The flow reattachment is about 0.043m down stream from the corner.

In the same way, the computed heat flux coefficient $(-\kappa \partial T/\partial n)$ is compared to experimental data. As expected the recirculation length is underestimated. However, the correct level of the heat flux coefficient after reattachment is found. Agreement is better with the two-layers approach.

Computed pitot pressure profiles, normalized by the free stream static pressure, through the boundary layer at $50mm$ and $100mm$ from the cone/sting corner are compared to experimental data in figures 10.32, 10.33.

The pitot pressure is computed from the local Mach number and static pressure by means of Rayleigh pitot formula:

$$\frac{p_t}{p_w} = (\frac{(\gamma+1)M^2}{2})^{\gamma/(\gamma-1)}(\frac{\gamma+1}{2\gamma M^2-(\gamma-1)})^{1/(\gamma-1)}$$

where $\gamma = 1.4$. p_w is the surface static pressure.
Moreover, we show the pressure, Mach and eddy viscosity contours in figures 10.34-10.36.

6.3 An Axisymmetric Flows with Swirl

In this section we show some results obtained with our Navier-Stokes solver using the two-layers technique for an axisymmetric flow where the third component (u_θ) of the velocity is non zero (see Appendix A10). The geometry corresponds to a high speed rotating mixer, fast enough to reach transonic speeds in the convergent part. This test case consists of a rotating internal flow in a convergent-divergent nozzle. The mesh for this case has 3951 nodes. The inflow Mach number is 0.4 and the Reynolds number is 10^6 . The inflow temperature in 290 K. At the inflow boundary a constant profile is assumed for all the variables. The solid body is adiabatic and the velocity and turbulent kinetic energy are zero at the wall.

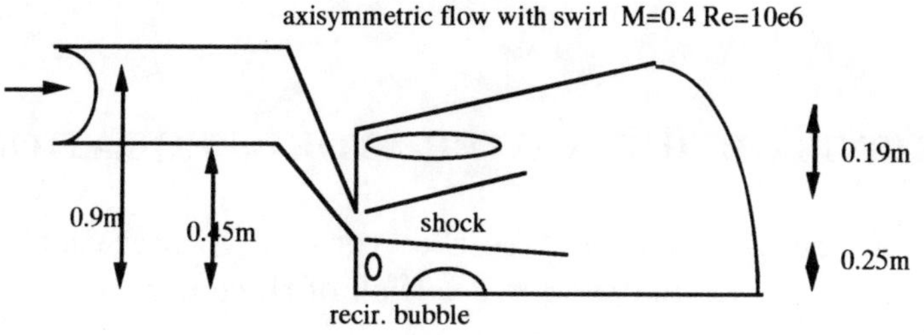

Figure 10.37

In simple axisymmetric situation without swirl, only one recirculation bubble is observed at the body side. When an inlet profile is assumed for u_θ, a new recirculation bubble appears centered on the symmetry line. This recirculation bubble appears for approximately $u_\theta = 0.3u_z$. The size of the bubble increases with the inlet amount of swirl. In the following pictures (10.38-10.42) we give some results obtained for $u_\theta = 0.5u_z$ at the inlet boundary. The pictures show respectively the velocity distribution, isolines for u_θ, the mach number, k and μ_{turb}. Lack of experimental results for this case makes the comparison and validation of these results difficult.

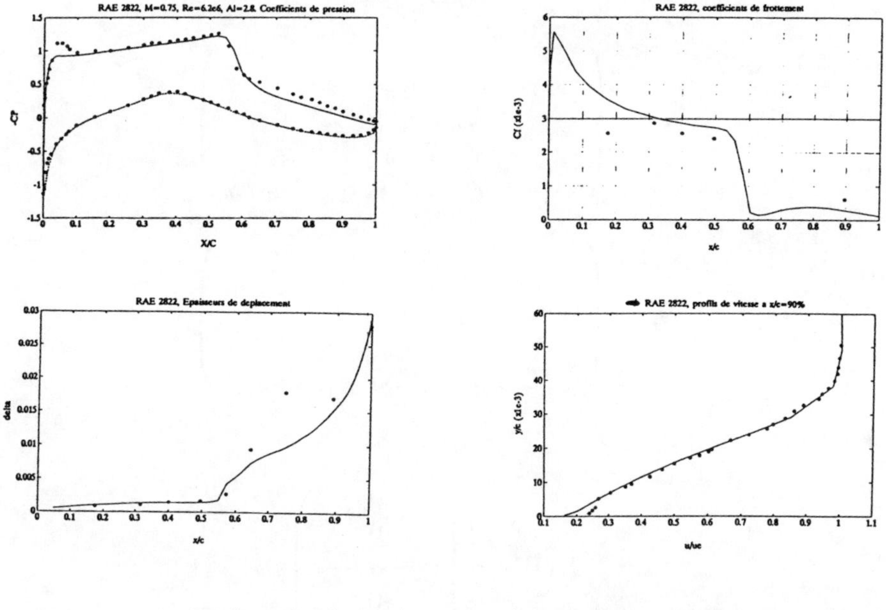

Figures 10.18-21 Cp (top left), Cf (top right), δ (bottom left) and velocity profile for a RAE2822 profile

AXISYMMETRIC NAVIER-STOKES EQUATIONS

Let $W = (\rho, \rho u_z, \rho u_r, \rho u_\theta, \rho E)$ be the vector of conservation variables in axisymmetric coordinates (z, r, θ). If all of the θ-derivatives are zero, the Navier-Stokes equations are:

$$\frac{\partial W}{\partial t} + \nabla.F(W) - \nabla.N(W) = H(W),$$

$$F_z = \begin{pmatrix} \rho u_z \\ \rho u_z^2 \\ \rho u_z u_r \\ \rho u_z u_\theta \\ (\rho E + p) u_z \end{pmatrix},$$

$$F_r = \begin{pmatrix} \rho u_r \\ \rho u_z u_r \\ \rho u_r^2 \\ \rho u_r u_\theta \\ (\rho E + p) u_r \end{pmatrix},$$

$$N_z = \begin{pmatrix} 0 \\ \tau_{zz} \\ \tau_{zr} \\ \tau_{z\theta} \\ \kappa \frac{\partial T}{\partial z} + u_z \tau_{zz} + u_r \tau_{zr} + u_\theta \tau_{z\theta} \end{pmatrix},$$

$$N_r = \begin{pmatrix} 0 \\ \tau_{rz} \\ \tau_{rr} \\ \tau_{r\theta} \\ \kappa \frac{\partial T}{\partial r} + u_z \tau_{zr} + u_r \tau_{rr} + u_\theta \tau_{r\theta} \end{pmatrix},$$

and

$$H(W) = \begin{pmatrix} 0 \\ 0 \\ \frac{\tau_{\theta\theta}}{r} + \frac{\rho u_\theta^2}{r} \\ 0 \\ 0 \end{pmatrix},$$

with :

$$\tau_{zz} = \mu(\frac{\partial u_z}{\partial z} - \frac{2}{3} \nabla .u), \quad \tau_{rz} = \mu(\frac{\partial u_r}{\partial z} + \frac{\partial u_z}{\partial r}),$$

$$\tau_{rr} = \mu(\frac{\partial u_r}{\partial r} - \frac{2}{3} \nabla .u), \quad \tau_{\theta\theta} = \mu(\frac{u_r}{r} - \frac{2}{3} \nabla .u),$$

$$\tau_{r\theta} = \mu(\frac{\partial w}{\partial r} - \frac{u_r}{r}), \quad \tau_{z\theta} = \mu\frac{\partial w}{\partial z},$$

where $\nabla = (\partial_z, \partial_r + \frac{1}{r})^T$

$$\nabla.u = \frac{\partial u_z}{\partial z} + \frac{\partial u_r}{\partial r} + \frac{u_r}{r}$$

$$p = (\gamma - 1)\rho T, E = T + \frac{1}{2}|u|^2,$$

$$\kappa = \frac{\mu\gamma}{Pr}, \gamma = 1.4, Pr = 0.72.$$

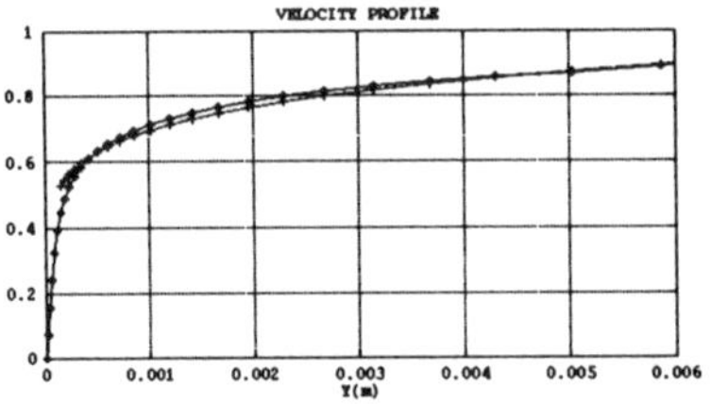

Figure 10.2. *Flow over a flat plate: computed u and the theoretical 1/7 power law.*

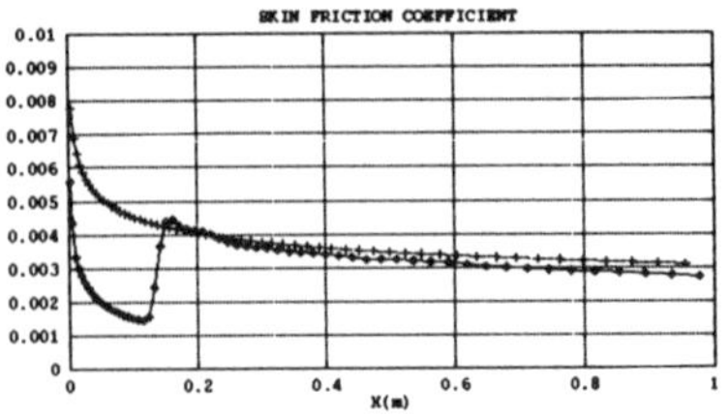

Figure 10.3. *Flow over a flat plate: computed and theoretical skin friction coefficient.*

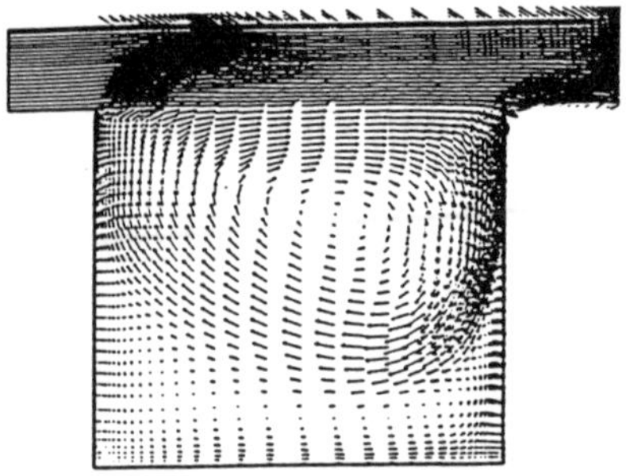

Figure 10.6. *Flow in a cavity: velocity field.*

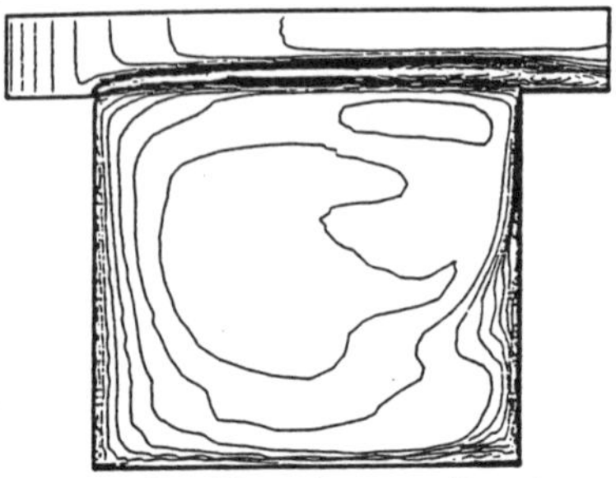

Figure 10.7. *Flow in a cavity: k contours.*

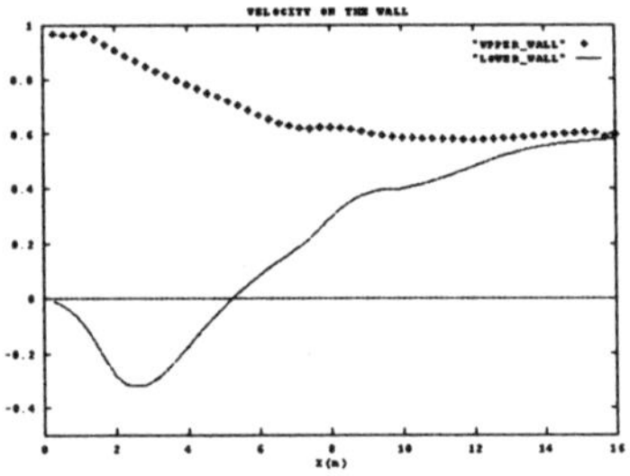

Figure 10.9. *Backward step: computed u on the walls.*

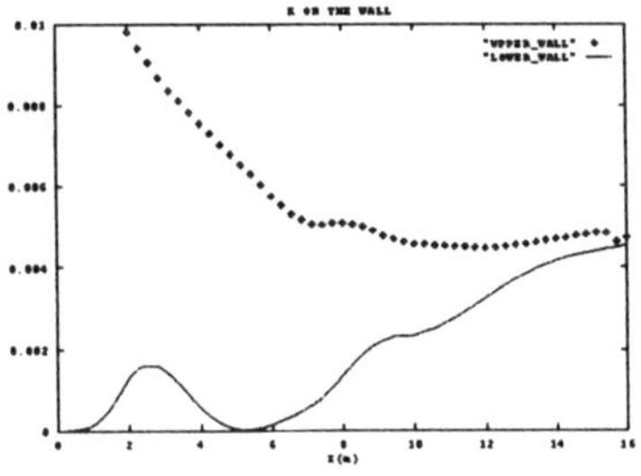

Figure 10.10. *Backward step: computed k on the upper and lower walls.*

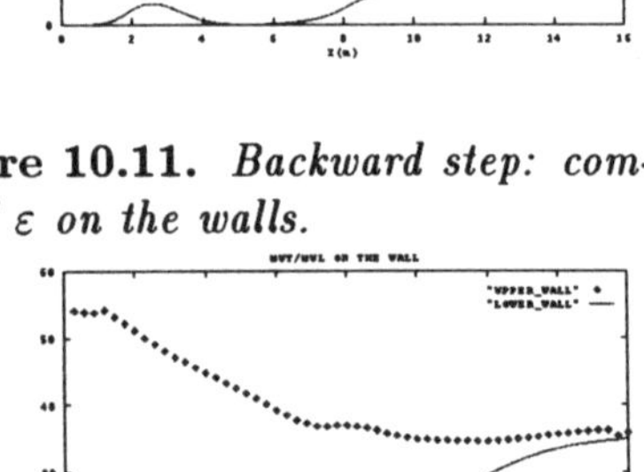

Figure 10.11. *Backward step: computed ε on the walls.*

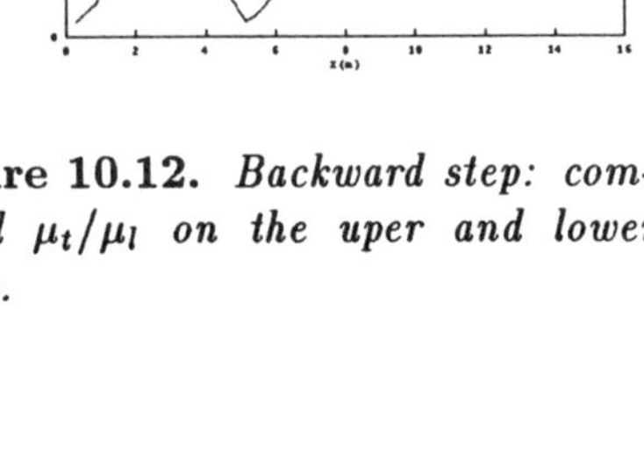

Figure 10.12. *Backward step: computed μ_t/μ_l on the uper and lower walls.*

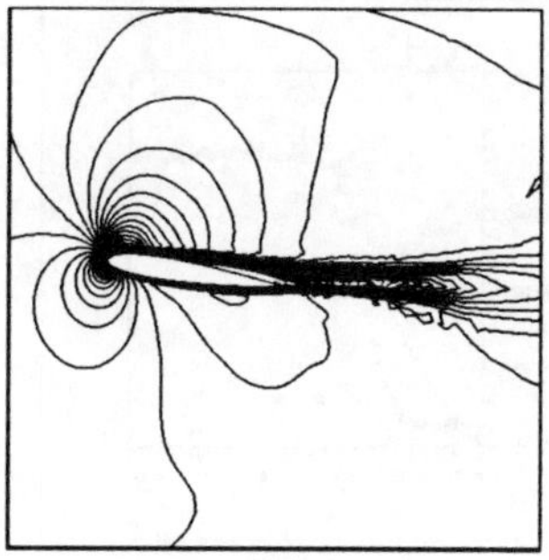

Figure 10.14. *Transonic airfoil: Mach number contours.*

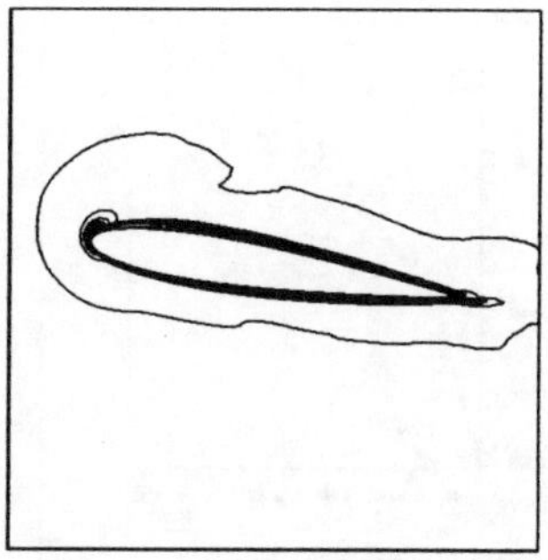

Figure 10.15. *Transonic airfoil: k contours.*

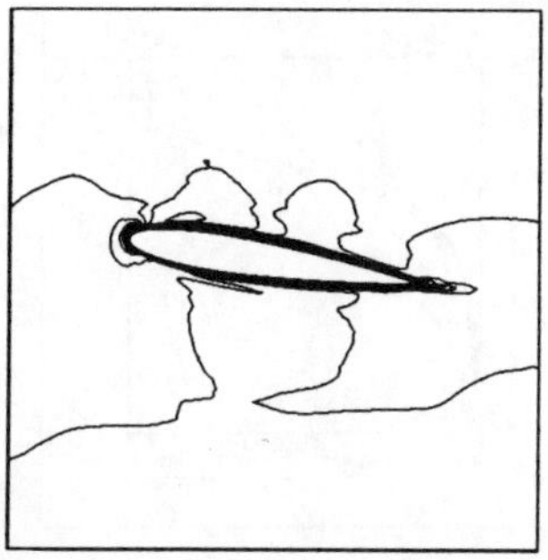

Figure 10.16. *Transonic airfoil: eddy viscosity contours.*

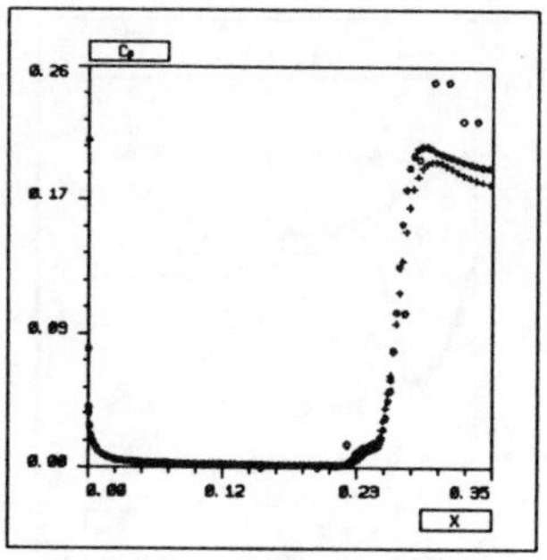

Figure 10.23. *Hypersonic corner: wall pressure (+ lam, * turb, o data).*

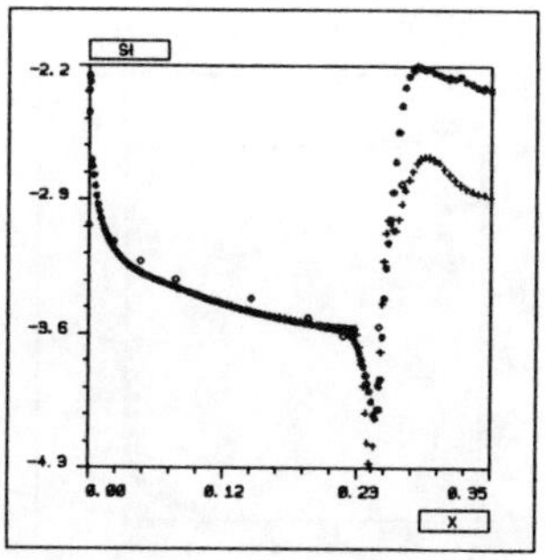

Figure 10.24. *Hypersonic corner: heat flux coefficient (+ lam, * turb, o data).*

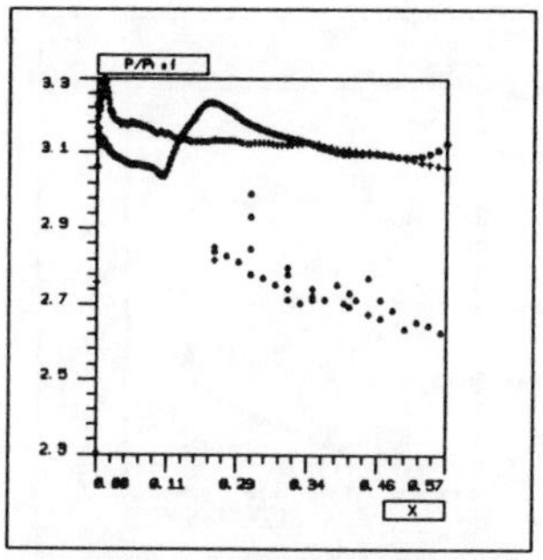

Figure 10.25. *Axisymetric cone: wall pressure(+ total, * trans, o data).*

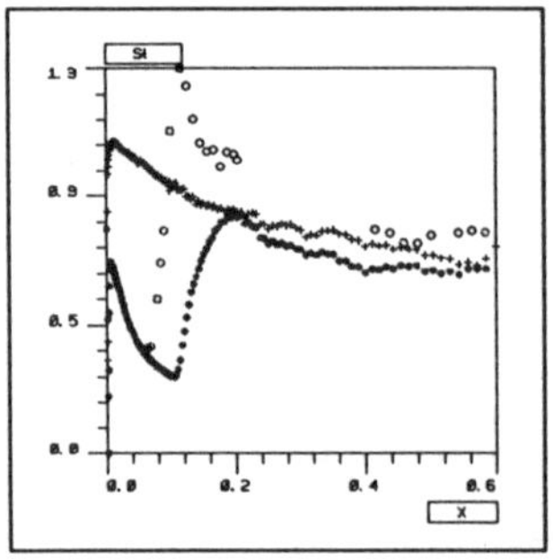

Figure 10.26. *Axisymetric cone: heat flux coef.(+ total, * trans, o data).*

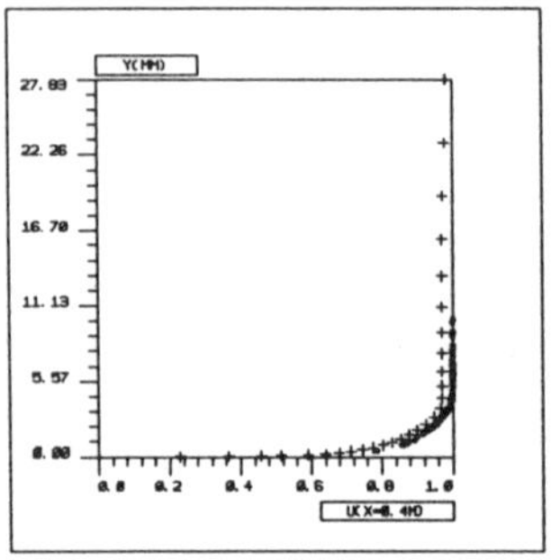

Figure 10.27. *Axisymetric cone: velocity profile (+ comp, * data).*

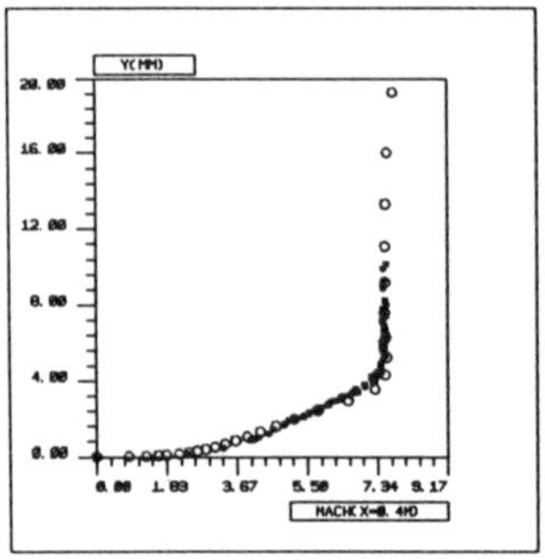

Figure 10.28. *Axisymetric cone: Mach number profile (o comp, * data).*

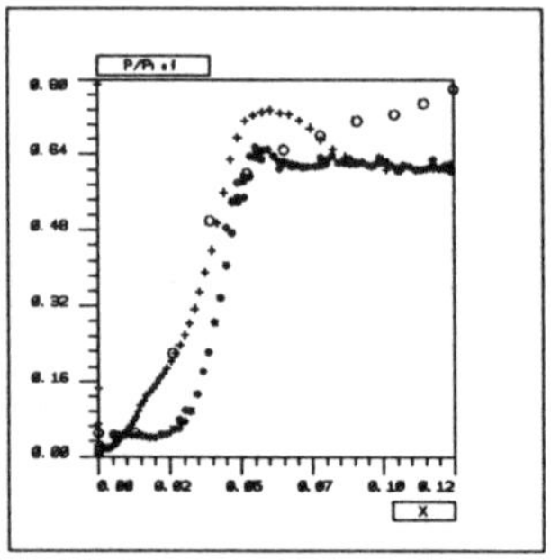

Figure 10.30. *Base flow: wall pressure (+ comp., * Hillier, o Netterfield).*

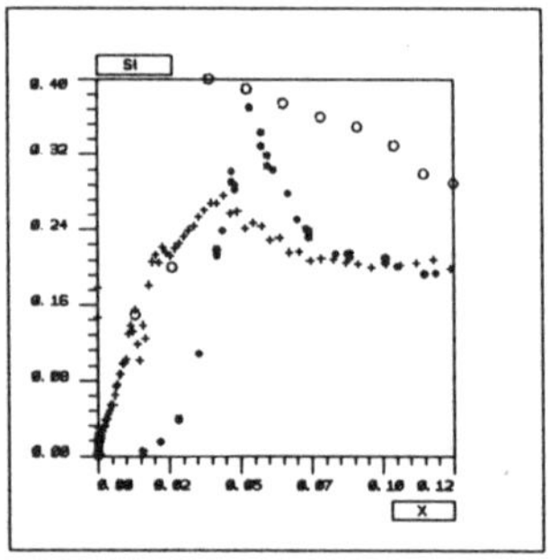

Figure 10.31. *Base flow: heat flux coef.(+ comp, * Hillier, o Netterfield).*

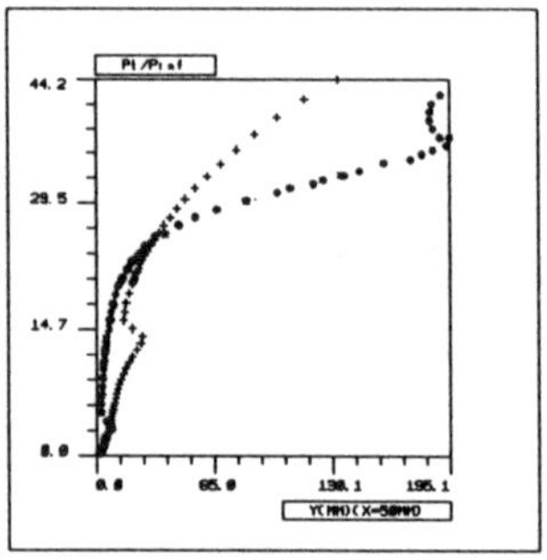

Figure 10.32. *Base flow: normalized pitot pressure at 50.mm (+ comp, * Hillier).*

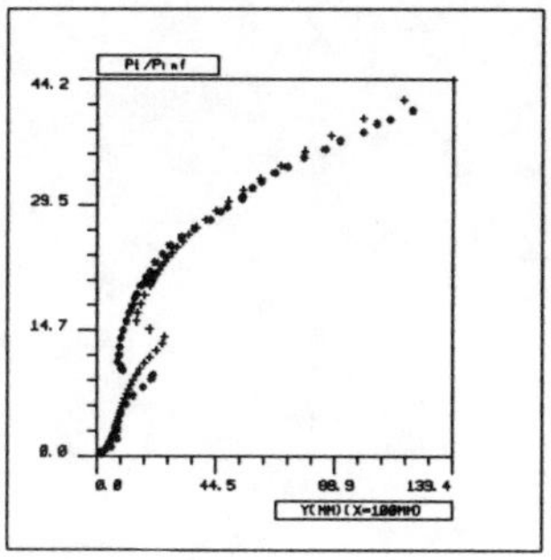

Figure 10.33. *Base flow: pitot pressure at 100.mm (+ comp, * Hillier).*

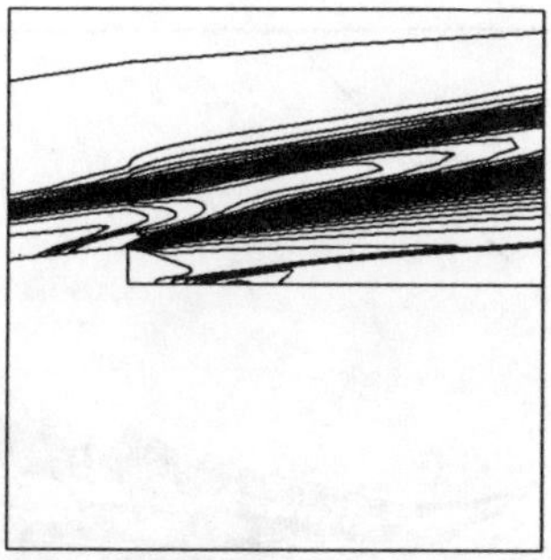

Figure 10.34. *Base flow: p/p_∞ contours.*

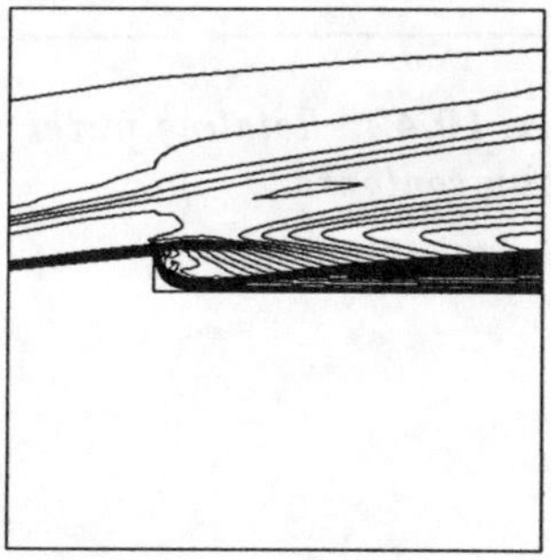

Figure 10.35. *Base flow: Mach number contours.*

Figure 10.36. *Base flow: eddy viscosity contours.*

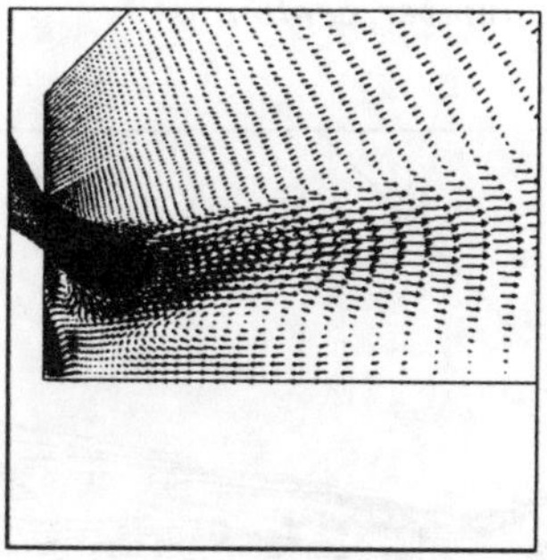

Figure 10.38. *Rotating mixer : Velocity field.*

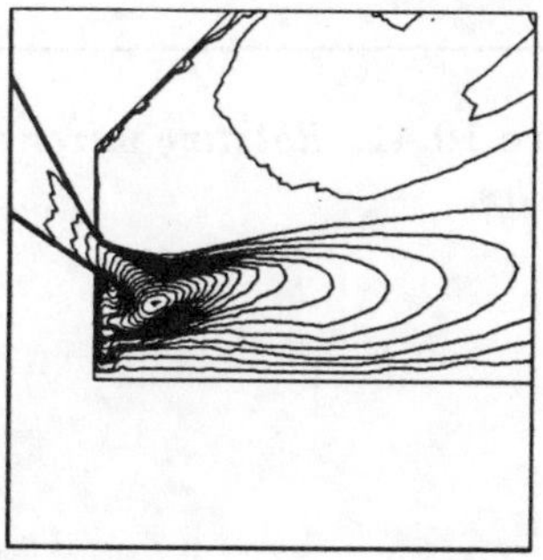

Figure 10.39. *Rotating mixer : iso v_θ contours.*

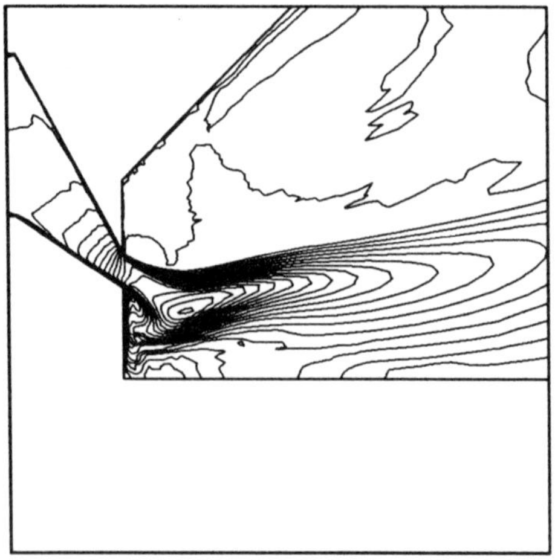

Figure 10.40. *Rotating mixer : Mach number contours.*

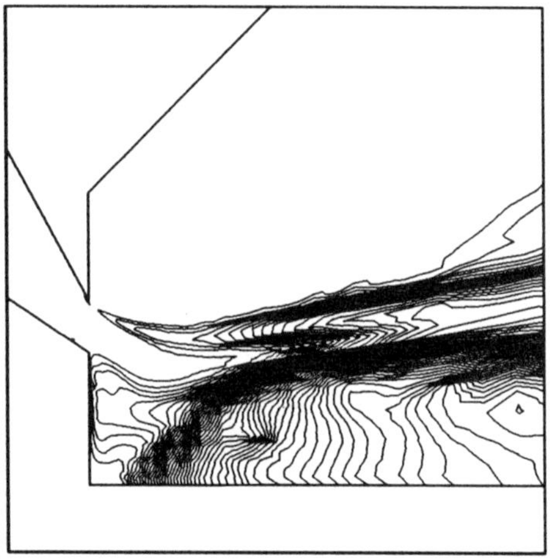

Figure 10.42. *Rotating mixer : eddy viscosity contours.*

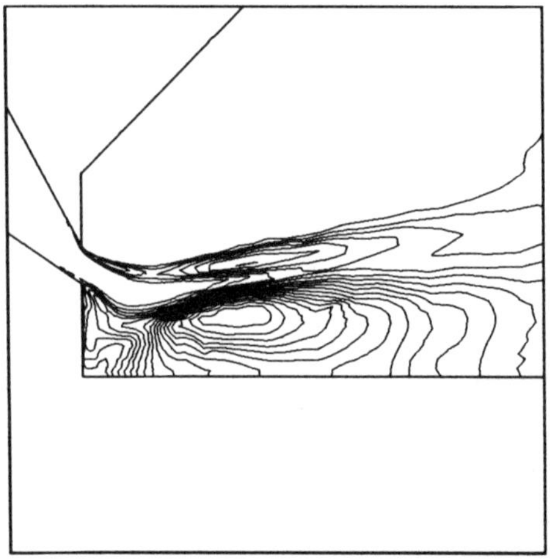

Figure 10.41. *Rotating mixer : iso k contours.*

CHAPTER 11

MATHEMATICAL METHODS FOR TURBULENT TRANSPORT

1. POSITION OF THE PROBLEM

It is clear from the previous chapters that everything relies on Reynolds' hypothesis: $\langle u' \otimes u' \rangle$ parallel to $\nabla u + \nabla u^T$. This hypothesis was shown to be a natural consequence of a weaker assumption: $\langle u' \otimes u' \rangle$ function of $\nabla u + \nabla u^T$. In the next two chapters, we shall investigate the mathematical tools which may lead to infer such an hypothesis or, more generally, lead to a theory of turbulence modeling and perhaps even to models similar to $k - \epsilon$.
In this chapter we shall concentrate on Hypothesis (H2) of Chapter 4 which was used systematically to replace turbulent transport by diffusion in establishing the $k - \epsilon$ model. We shall proceed in two steps: first we shall analyse the much easier problem of passive transport by a "turbulent" velocity field. The full Navier-Stokes equations will be analysed in the next chapter, with one of the tools presented here: homogenization.

The problem of passive transport by a "turbulent" velocity field is as follows:
Let Ω be an open set of R^n with boundary Γ.
Let u be a given divergence free 'turbulent' velocity field and define

$$\Gamma^- = \{x \in \Gamma : u(x).n(x) < 0\}$$

Let $\langle\cdot\rangle$ be a filter and let c be a passive scalar governed by a convection equation:

$$\partial_t c + u\nabla c = f, \quad \text{in } \Omega\times]0,T[, \text{ with } c(0) = c_0, \quad c|_{\Gamma^-} = 0.$$

Then find a computational method for $C = \langle c\rangle$?

For simplicity we assume that $\nabla.u = 0$ in $\Omega\times]0,T[$ and $u.n = 0$ at the boundary $\Gamma = \partial\Omega$.

This is a first step toward turbulence modeling. Indeed averaging this equation is as difficult as for the Navier-Stokes equations. For example with c=C+c' and u=U+u', Reynolds analysis gives:

$$\partial_t(C + c') + (C + u')\nabla(C + c') = F + f',$$

so

$$\partial_t C + U\nabla C + \nabla.\langle u'c'\rangle = F.$$

The closure problem here comes down to find a formula for $\langle u'c'\rangle$. Several methods are at our disposal. First the probabilistic approach, then Homogenizations theory and finally the old method of multiple scale expansion.

2. THE PROBABILISTIC APPROACH

Let us replace the problem by a stochastic PDE by considering u random: $u(x,t,\omega)$ is now a given div-free random velocity field with $u.n = 0$ on Γ. Then the following is true :

Proposition (Kesten-Papanicolaou[1980]). *If u is stationary, independent of t and mixing* [1] *and its mean, $U = \langle u\rangle$, is large compared with $\sqrt{|u'|^2}$ where u'=U-u , then $C = \langle c\rangle$, satisfies*

$$\partial_t C + U\nabla C - \nabla.[M\nabla C] = F$$

[1] A random vector field w is mixing if, for any two sets of points $\{x^i\}_1^m, \{y^j\}_1^n$ with $||x^i - y^j|| > d, \forall i,j$, and any functions $F(w(x^1),...,w(x^m)), G(w(y^1),...,w(y^n))$ there exists $\alpha(d)$ decaying sufficiently fast to zero with d such that : $|\langle FG\rangle - \langle F\rangle\langle G\rangle| \leq \alpha(d)|F|_\infty|G|_\infty$ independently of m, n.

where the matrix M is function of the statistics of u and given by

$$M_{ij} = \frac{1}{2}\int_{-\infty}^{+\infty} \langle u_i'(x+Ut)u_j'(x)\rangle dt$$

It is conjectured that the result holds also, in three dimensions, and also when U and u' are of the same order; however the formula for M is much more complicated. There are counter examples to this results when u is not mixing or has zero mean (Childress[1989], Avelaneda et al[1992]). The theory can be extended to the convection-diffusion equation but extensions to nonlinear cases like Navier-Stokes equations is an open problem , in general (see Kesten et al [1980], Isichenko[1992], Hou[1991], Maslov[1987]).

3. PROPAGATION OF OSCILLATIONS

Another approach is to use the tools of functional analysis to study limits with respect to small parameters in the problem, such as the inverse of the Reynolds number, ν, or the mesh size h of the numerical method and/or the characteristic length scale of the filter.
An important case, known as propagation of oscillations (diPerna[1987], Serre[1991]) is when the small parameter is in the initial conditions.

Denote the small parameter by ϵ and a given velocity field u^ϵ; consider

$$\partial_t c^\epsilon + u^\epsilon \nabla c^\epsilon = f \quad \text{in} \quad \Omega\times]0,T[\quad c^\epsilon(0) = c_0^\epsilon \quad \text{in } \Omega \quad c^\epsilon_{|\Gamma^-} = 0$$

Then the problem is to find $C = \lim_{\epsilon\to 0} c^\epsilon$.
When the convergence $c_0^\epsilon \to c^0$ is strong, nothing happens, as we shall see; but when it is weak, as in the case $c_0^\epsilon = sin(x/\epsilon)$ the problem is difficult.

3.1 Strong convergence

If $c_0^\epsilon \to c_0$ strongly in L^2 and $u^\epsilon \to U$ strongly in $H(div,\Omega)$, the problem is easy because the PDE gives a bound on c^ϵ . Indeed by multiplying it by c^ϵ, it gives

$$\frac{1}{2}|c^\epsilon(t)|_0^2 \le \int_0^T |f|_0|c^\epsilon|_0 + \frac{1}{2}|c_0^\epsilon|_0^2$$

So c^ϵ is bounded in $L^\infty(0,T;L^2)$; there exists a subsequence for which $c^\epsilon \to C$ weakly in L^2 and one can pass to the limit in the variational formulation of the PDE:

$$\int_{\Omega\times]0,T[} (c^\epsilon \partial_t w + c^\epsilon u^\epsilon \nabla w + fw) - \int_\Gamma c^\epsilon(x,0)w(x,0) = 0$$

for all $w \in L^\infty(0,T;H^1)$ with $w(T)=0$. Indeed $u^\epsilon c^\epsilon \to UC$ in L^1. So C satisfies

$$\partial_t C + U\nabla C = f.$$

3.2 Weak convergence

When the flow is "turbulent" we may not know that $u^\epsilon \to U$ strongly in L^2 even. If it converges weakly only then of course $c^\epsilon u^\epsilon$ may not converge to CU. It is usually necessary to make some assumption on u^ϵ to be able to conclude. In section 4 we shall assume that $u^\epsilon(x,t) = u(x,t,x/\epsilon)$ is periodic with respect to the last variable.

4. HOMOGENIZATION

When oscillations are periodic or quasi-periodic and bounded in L^2, the problem of finding an equation for the mean of the solution of a convection equation can be approached with tools of functional analysis such as compensated compactness (Tartar[1989], Murat[1978]) Young measures (Diperna[1983]) and H-measures (Tartar[1987]); but the results can be derived formally also by multiple-scales asymptotic expansions as we shall see in the next paragraph.

4.1 L^2 bounded oscillations; the general case

Consider the linear problem

$$\partial_t c^\epsilon + u^\epsilon \nabla c^\epsilon = f, \quad c^\epsilon(x,0) = c_0^\epsilon(x), \quad \text{in} R^d,$$

where $u^\epsilon(x) = u(x,y)|_{y=x/\epsilon}$ and $c_0^\epsilon(x) = c_0(x,y)|_{y=x/\epsilon}$ are periodic functions in y over a parallelogram Y of R^d and such that:

$$\nabla_x . u = 0 \quad \nabla_y . u = 0.$$

Lemma (Nguetseng-E[1987])
If $\{\phi^\epsilon\}_\epsilon$ is bounded in $L^\infty(R^+, L^2_{loc}(R^d))$ then for a subsequence there exists $\Phi \in L^2_{loc}(R^d \times R^+ \times Y)$ such that, when $\epsilon \to 0$,

$$\int_{R^d\times R^+} \phi^\epsilon(x,t)\omega(x,t,\frac{x}{\epsilon})dxdt \to \int_{R^d\times R^+\times Y} \Phi(x,t,y)\omega(x,t,y)dxdtdy,$$

for all smooth Y-periodic functions ω with compact (x,t)-support.

Theorem
In the limit $\epsilon \to 0$, the weak limit c of c^ϵ, together with Φ is solution of

$$\partial_t c + u\nabla_x c + u\nabla_y \Phi = f, \quad c(x,0) = c_0(x,y),$$
$$u\nabla_y c = 0 \qquad \forall x, y, t,$$

where Φ is Y-periodic.

Remark
By increasing the dimension of the space from x to x, y the oscillatory problem is transformed into a non-oscillatory PDE, but the result is simple only at first sight.
Once $c(x,y,t)$ is computed then all moments are easy to compute by application of the Lemma.

Proof of the theorem
In variational form the problem is

$$\int_{R^d\times R^+} [c^\epsilon(\partial_t\omega^\epsilon + u^\epsilon\nabla\omega^\epsilon) + f\omega^\epsilon] - \int_{R^d} c^\epsilon(x,0)\omega^\epsilon(x,0) = \quad 0,$$

for all ω^ϵ, smooth with compact support.
Since c^ϵ is bounded in $L^\infty(0,T;L^2(R^d))$ (see §3.1) we can apply the lemma with ϕ^ϵ replaced by c^ϵ and Φ by c(x,t,y). Choose $\omega^\epsilon = \epsilon\omega'(x,t,x/\epsilon)$ where ω' is Y-periodic. Then the variational form yields

$$\int_{R^d\times R^+\times Y} (u\nabla_y\omega')c(x,t,y)dxdtdy = 0 \quad \text{i.e.} \quad u\nabla_y c = 0$$

Next take $\omega^\epsilon = \omega(x,t,x/\epsilon)$ with $u\nabla_y\omega = 0$; then by the same lemma ($\partial_t\omega^\epsilon + u^\epsilon\nabla_x\omega^\epsilon$ is now $\frac{x}{\epsilon}$ -periodic):

$$\int_{R^d\times R^+} c^\epsilon[\partial_t\omega^\epsilon + u^\epsilon\nabla\omega^\epsilon] = \int_{R^d\times R^+} c^\epsilon[\partial_t\omega^\epsilon + u^\epsilon\nabla_x\omega^\epsilon]$$
$$\to \int_{R^d\times R^+\times Y} c[\partial_t\omega + u(x,t,y)\nabla_x\omega].$$

So

$$\int_{R^d\times R^+\times Y} c[\partial_t\omega + u\nabla_x\omega] - \int_{R^d\times Y} c(x,0,y)\omega(x,0) = -\int_{R^d\times R^+} f\omega$$

for all ω Y-periodic satisfying $u\nabla_y\omega = 0$.

Notice now that

$$\int \phi\omega = 0 \quad \forall\omega \text{ with } u\nabla_y\omega = 0 \quad \Leftrightarrow \quad \int(\phi + u\nabla_y\Phi)\omega = 0 \quad \forall\omega.$$

So for some Φ, Y-periodic,

$$\partial_t c + u\nabla_x c + u\nabla_y\Phi = f, \quad u\nabla_y c = 0.$$

4.2 Applications

Definition

A flow field $u(x,y)$ is ergodic in y, i.e. if for all x,y, the solution of

$$\frac{dX}{dt} = u(x,X,t), \quad X(0) = y,$$

fills Y when $t\to\infty$. The curve $t\to X$ is called a streamline.

Notice that $u\nabla c = 0$ in Y implies that c is independent of y when u is ergodic, because c is equal to its initial value everywhere.

Corollary

If u is ergodic then $\nabla_y.[uc] = 0 \Rightarrow c = c(x,t)$ and so by averaging in y the equation of c we find that $C = \langle c\rangle_y$ satisfies :

$$\partial_t C + U\nabla_x C = \langle f\rangle_y.$$

Thus no diffusion term appears in the equation for the mean in this case.

An example with integro-differential diffusion

Take $u = (u_1(y_2), 0)^T$ and c^0 independent of ϵ. Obviously $\nabla.u = 0$ but u is not ergodic. The theorem tells us that

$$\partial_t c + u_1 \partial_{x_1} c + u_1 \partial_{y_1} \Phi = 0 \quad u_1 \partial_{y_1} c = 0$$

so c does not depend upon y_1 and by averaging the first equation in y_1 we obtain

$$\partial_t c + u_1(y_2) \partial_{x_1} c(x, y_2) = f \quad c(0, x, y_2) = c^0(x).$$

The solution is

$$c(x, t, y_2) = \langle c^0(x - u_1(y_2)t) \rangle$$

But if one wants a convection diffusion equation for the mean $C = \langle c \rangle$ in y_2 of c then one must express $\partial_t C + \langle u_1 \rangle \partial_{x_1} C$ in terms of $\partial_{ij} C$. Amirat et al[1989] found that for some family of measures σ_{x_2}:

$$\partial_t C + \langle u_1 \rangle \partial_{x_1} C - \int_0^t \int_{u_{min}}^{u_{max}} \partial_{x_1}^2 C(s, x_1 - \lambda(t-s), x_2) d\sigma_{x_2}(\lambda) ds = 0$$

Other examples can be found in Childress[1989],Hou[1992],Weinan-E[1991].

4.3 Euler Equations in 1D

In one dimension Serre[1991] extended the method and solved the problem completely in absence of viscosity. We present his result without proof.

Euler equations in one dimension may be written as

$$(\partial_t + u\partial_x)\rho + \rho \partial_x u = 0$$
$$(\partial_t + u\partial_x)u + \frac{1}{\rho}\partial_x p = 0$$
$$(\partial_t + u\partial_x)e + \frac{1}{\rho} p \partial_x u = 0,$$

where $p = (\gamma - 1)\rho e$. The system can be rewritten in terms of entropy variables, i.e.

$$(\partial_t + u\partial_x)u + \frac{1}{\rho}\partial_x p = 0$$
$$(\partial_t + u\partial_x)p + \rho c^2 \partial_x u = 0$$
$$(\partial_t + u\partial_x)S = 0,$$

where $c^2 = \gamma(\gamma - 1)e$ and $S = \log(p\rho^{-\gamma})$. The system is integrated on $R\times]0, T'[$ with initial conditions

$$\rho(0, x) = \rho^0(x, y), \quad u(0, x) = u^0(x, y), \quad e(0, x) = e^0(x, y),$$

where y is a parameter. The idea is to solve the system for all y instead of just $y = x/\epsilon$. Then the solution depends upon y and the problem is to find a system of equations for $\bar{u}, \bar{p}, \bar{S}$, where

$$\bar{u} = \lim_{Y\to\infty} \frac{1}{2Y} \int_{-Y}^{Y} u(x, y)dy$$

Under certain conditions (no shocks, integrability...) the answer is

$$(\partial_t + \bar{u}\partial_x)\bar{u} + \frac{1}{\langle\rho\rangle}\partial_x \bar{p} = 0$$
$$(\partial_t + \bar{u}\partial_x)\bar{p} + \frac{1}{\langle\rho^{-1}c^{-2}\rangle}\partial_x \bar{u} = 0$$
$$\partial_t\langle\rho\varphi(S)\rangle + \partial_x(\bar{u}\langle\rho\varphi(S)\rangle) = 0 \quad \forall \ \varphi.$$

The filter $\langle\cdot\rangle$ is defined on functions of p, S by

$$\langle f\rangle = \lim_{Y\to\infty} \frac{1}{2Y} \int_{-Y}^{Y} f(\bar{p}, S(y))dy.$$

Hence, with $\rho = p^{1/\gamma}\, e^{\frac{S}{\gamma}}$

$$\langle\rho\rangle = \lim_{Y\to\infty} \frac{1}{2Y} \int_{-Y}^{Y} \bar{p}^{\frac{1}{\gamma}} e^{\frac{S(y)}{\gamma}} dy,$$

and similarly for $\rho^{-1}c^{-2}$ and $\rho\varphi(S)$, by the formulae:

$$\rho\varphi(S) = \varphi(S)p^{\frac{1}{\gamma}} e^{\frac{S}{\gamma}}, \quad \frac{1}{\rho c^2} = \frac{1}{\gamma p}.$$

It is an integro-differential system which must be analysed further for practical applications. In particular when the initial conditions are quasi-periodic in y, the quantities $\langle\rho\rangle$ must be evaluated. Extension of the method to several space dimensions and to the Navier-Stokes equations can be found in Serre[1990][1991].

Remark

diPerna et al[1985] analysed Euler equations with oscillations in the initial conditions periodic and of small amplitudes. Oscillations are shown to propagate according to laws similar to geometrical optic.

5. HOMOGENIZATION BY MULTIPLE SCALE EXPANSIONS

5.1 The ergodic case with large amplitudes

Consider the case of a velocity field of the form

$$u^\epsilon(x,t) = U(x,t) + \frac{1}{\epsilon}u(x,t,\frac{x}{\epsilon})$$

where u is periodic with respect to its last argument, $y = x/\epsilon$, in $Y =]0,1[^d$, has zero y-mean, and is ergodic. As before we assume U and u divergence free in x and y and $U.n = u.n = 0$ on Γ.

The problem considered is again

$$\partial_t c^\epsilon + (U + \frac{1}{\epsilon}u^\epsilon)\nabla c^\epsilon = f$$

with given periodic initial data.

Proposition

For smooth data $C(x,t) = \lim_{\epsilon\to 0}|Y|^{-1}\int_Y c^\epsilon(x,t,y)dy$ *satisfies*

$$\partial_t C + \nabla_x.(UC) - \nabla_x.[\langle uA\rangle\nabla_x C] = 0.$$

where the tensor A *is solution of*

$$u\nabla_y A = u \quad \textit{with } A \textit{ Y-periodic}$$

and $\langle uA\rangle = |Y|^{-1}\int_Y u(x,t,y)A(x,t,y)dy$.

Proof

One may search for c^ϵ in the following form (multiple scales expansion, see Bensoussan et al[1978])

$$c^\epsilon = c^0(x,t,y) + \epsilon c^1(x,t,y) + \epsilon^2 c^2(x,t,y) + ...|_{y=\frac{x}{\epsilon}},$$

where the c^i are Y-periodic.

By the rules of differentiation we have

$$\nabla c^\epsilon = [\frac{1}{\epsilon}\nabla_y c + \nabla_x c]|_{y=\frac{x}{\epsilon}}.$$

So the equation for c^ϵ becomes

$$[\partial_t + (U(x,t) + \frac{1}{\epsilon}u(x,t,y))(\frac{1}{\epsilon}\nabla_y + \nabla_x)]$$
$$[c^0(x,t,y) + \epsilon c^1(x,t,y) + ...]|_{y=\frac{x}{\epsilon}} = f.$$

The left hand side is

$$\begin{aligned}&1/\epsilon^2\nabla_y.(uc^0)\\ &+\frac{1}{\epsilon}[\nabla_y.(uc^1) + \nabla_x.(uc^0)]\\ &+\partial_t c^0 + \nabla_x.(Uc^0) + \nabla_x.(uc^1) + \nabla_y.(uc^2)\\ &+\epsilon[\partial_t c^1] + ...\end{aligned}$$

Set all powers of ϵ to zero. The ϵ^{-2} power gives

$$u\nabla_y.c^0 = 0.$$

When the flow is ergodic, this equation implies c^0 independent of y.

The next term is

$$\nabla_y.(uc^1) + \nabla_x.(uc^0) = 0.$$

This means that $c^1 = -A\nabla_x c^0$ where A is the solution of

$$u\nabla_y A = u \quad \text{with } A \ \ Y\text{-periodic.}$$

The third term gives

$$\nabla_y.(uc^2) + \partial_t c^0 + \nabla_x.(Uc^0) + \nabla_x.(uc^1) = 0.$$

This gives an equation for the mean $C = \langle c^0 \rangle$. Indeed, by taking the mean of this equation we find

$$\partial_t C + \nabla_x.(UC) - \nabla_x.[\langle uA \rangle \nabla_x C] = 0 \text{ in } \Omega \times]0, T[.$$

$$u\nabla_y A = u \quad A \text{ Y-periodic.}$$

Thus turbulent transport with large amplitude leads to to convection and diffusion for the mean in this case.

5.2 The Convection Diffusion equation with Bounded Oscillations

Consider the problem

$$\partial_t c^\epsilon + u^\epsilon \nabla c^\epsilon - \epsilon^2 \nu \Delta c^\epsilon = f,$$

with Dirichlet boundary conditions independent of ϵ and initial conditions $c_0(x, y)|_{y=x/\epsilon}$, periodic in Y. Furthermore assume that

$$u^\epsilon(x, t) = u(x, t, y)|_{y=\frac{x}{\epsilon}} \quad Y - \text{periodic},$$

with u ergodic and divergence free both in x and y. We denote by $\langle u \rangle$ the mean of u with respect to $y = x/\epsilon$ over Y. Most results of the previous paragraphs extends to this case. However the following result defines a *corrector* and is more useful:

Proposition .
If u is ergodic in y then $\langle c^\epsilon - z^\epsilon \rangle$ is $o(\varepsilon)$ where z^ϵ is the solution of

$$\partial_t z^\epsilon + \langle u \rangle \nabla z^\epsilon - \epsilon \nabla_x.[\langle (u - \langle u \rangle) A \rangle \nabla_x z^\epsilon] = f,$$

where A is the solution of

$$u\nabla_y A = u - \langle u \rangle, \quad A \; Y - \textit{periodic}.$$

Remark

We have replaced a PDE with oscillatory coefficients and $0(\epsilon^2)$ diffusion by one with smooth coefficients and $0(\epsilon)$ diffusion. Here the diffusion effect of the L^2- bounded velocity appears as a corrector of order one. In more complex situations anomalous diffusion with coefficients different from ϵ can appear (see Fannjiang et al [1993], for example).

Proof of Proposition
Again search for c^ϵ as

$$c^\epsilon = c^0(x,t,y) + \epsilon c^1(x,t,y) + \epsilon^2 c^2(x,t,y) + ...|_{y=\frac{x}{\epsilon}},$$

where the c^i are Y-periodic. Recall that

$$\nabla c^\epsilon = [\frac{1}{\epsilon}\nabla_y c + \nabla_x c]|_{y=\frac{x}{\epsilon}}.$$

Hence

$$\begin{aligned} f =& [\partial_t + (\frac{1}{\epsilon}\nabla_y + \nabla_x).(u(x,t,y))] \\ & [c^0(x,t,y) + \epsilon c^1(x,t,y) + \epsilon^2 c^2(x,t,y) + ...]|_{y=\frac{x}{\epsilon}} \\ & = 1/\epsilon \nabla_y.(uc^0) + [\nabla_y.(uc^1) + \nabla_x.(uc^0)] + \partial_t c^0 \\ & + \epsilon[\nabla_x.(uc^1) + \nabla_y.(uc^2) + \partial_t c^1] + ... \end{aligned}$$

Then set all powers of ϵ to zero as before:

$$u\nabla_y c^0 = 0 \quad \Rightarrow \quad c^0 \text{ independent of y} \quad .$$

For the ϵ^0 term first take the mean in y and get

$$\partial_t c^0 + \langle u \rangle \nabla_x c^0 = f.$$

Next subtract what was found:

$$\nabla_y.(uc^1) + \nabla_x.([u - \langle u \rangle]c^0) = 0 \quad \Rightarrow \quad c^1 = -A\nabla_x c^0,$$

where A is a solution of

$$u\nabla_y A = u - \langle u \rangle \quad A \quad Y - periodic.$$

The third term is treated in the same way; first average in y:

$$\partial_t \langle c^1 \rangle + \nabla_x.\langle uc^1 \rangle = 0.$$

Now form

$$\begin{aligned} (\partial_t + & \langle u \rangle \nabla)[c^0 + \epsilon \langle c^1 \rangle] \\ & = \partial_t c^0 + \langle u \rangle \nabla_x c^0 + \epsilon[\partial_t \langle c^1 \rangle + \langle u \rangle \nabla_x.\langle c^1 \rangle] \\ & = f + \epsilon \nabla_x.[\langle (u - \langle u \rangle) A \rangle \nabla_x c^0]. \end{aligned}$$

5.3 Quasi-periodic functions

One of the problem of the previous analysis for turbulent flows is that we may approximate u by a function with multiple scales, but there are no obvious period if we want to approximate u by a periodic function. Fortunately the previous analysis extends to quasi-periodic functions (see Corduneanu[1968]) for which the mean is defined by

$$\langle u\rangle(x,t) = \lim_{r\to\infty} \frac{\int_{B(z,r)} u(x,t,y)dy}{\int_{B(z,r)} dy.}$$

Roughly speaking, almost-periodic functions have discrete Fourier spectrum, i.e. if

$$f(y) = \sum_{k\in Q} e^{ik.x} f_k \qquad Q \quad \text{a finite subset of the rational numbers}$$

then f is a quasi-periodic function [2]

6. CONCLUSION

Thus in most cases convection by an oscillatory velocity field has a diffusive effect on the mean; there are however many counter examples, primarily when the streamlines are closed.
The general theory involves compensated compactness and H and Young's measures. Serre[1991] and Weinan-E[1991] have shown that it is possible to use these tools to find effective equations for the means with the Navier-Stokes or Euler equations. However one may not be able to characterize sufficiently the results for practical uses unless strong assumptions are made on the scales in the velocity field; periodicity and quasi-periodicity are such good assumptions (homogenization).
There is no doubt that this theory will yield good turbulence models in the future. But already some insides to the problem can be found with simple multiple scales expansions. Multiple-scale expansions are less general but they form an easier working tool. We shall use it in the next chapter to find a turbulence model which bears some similarities with the $k-\epsilon$ model.

[2] The space of almost periodic functions is the closure of all trigonometric polynomials in the uniform norm. It can be shown that the mean defined above is invariant by change under translations and rotations.

CHAPTER 12

DERIVATION OF A TWO-EQUATIONS MODEL BY MULTIPLE-SCALES EXPANSION [1]

1. POSITION OF THE PROBLEM

We have seen in the previous chapter that multiple scales expansion is a simple tool to obtain equations for the means when there are periodic or quasi-periodic oscillations. Here we shall derive a turbulence model by the same method. The model is not $k-\epsilon$ but it bears some similarities with it.

Consider the Navier-Stokes equation with a small parameter in the viscosity and in the initial conditions.

$$\begin{aligned} &\partial_t u^\epsilon + (u^\epsilon \cdot \nabla)u^\epsilon - \epsilon^2 \mu \Delta u^\epsilon + \nabla p^\epsilon = 0, \\ &\nabla \cdot u^\epsilon = 0 \quad \text{in } R^3 \times\,]0,T[, \end{aligned} \tag{1}$$

$$u(x,0) = u_0(x) + \epsilon^{\frac{1}{3}} w_0(x, \frac{x}{\epsilon}) \text{ in } R^3. \tag{2}$$

where $w_0(x,y)$ is almost periodic in y (see §XI.5.3) and has zero y-mean. Here ϵ is a length scale and not the rate of turbulent energy dissipation which will be called e in this analysis; the turbulent kinetic energy k will also be called q and h will denote helicity.

The choice $\alpha = 2$ for the vanishing viscosity $\nu = \mu\epsilon^\alpha$ is based on Kolmogorov's scales.

[1] This chapter is based on Chacon et al[1993] and jointly written with T. Chacon, D Franco and F. Ortegon. It extends and sumarizes a research initiated in Perrier et al[1980], Papanicolaou et al [1981], MacLaughin et al [1982] and pursued in Begue[1983], Chacon[1986][1988], Ortegon[1989]

At time zero we have

$$q_0 = \epsilon^{\frac{2}{3}}\langle |w_0|^2\rangle \quad e_0 = \mu\epsilon^{\alpha-2+\frac{2}{3}}\langle|\nabla_y \times w_0|^2\rangle.$$

After Fourier transform the $K^{-\frac{5}{3}}$ law becomes (Lesieur[1987])

$$\langle u(x)u(x+r)\rangle \sim r^{\frac{2}{3}}$$

So the initial condition (2) is dimensionally compatible with Kolmogorov's range. However for ϵ to be in the Kolmogorov range, $\mu\epsilon^\alpha$ must be small compared with $\epsilon^{4/3}$ So the right choice is $\alpha > 4/3$.
We choose $\alpha = 2$ thus placing the analysis as a study of the interaction of large eddies with small eddies well within the Kolmogorov range instead of at the end of the inertial range.

2. RESULTS

Just as in the proposition of §XI.5.2, it will be found that $\{u^\epsilon, p^\epsilon\}$ are close to the solution $\{u, p\}$ of

$$\begin{aligned}
&\partial_t u + (u\cdot\nabla)u + \nabla p + \epsilon^{\frac{2}{3}}\nabla\cdot(q\mathbf{R}^*(C)) = 0,\\
&\nabla\cdot u = 0\\
&\partial_t a + (u\cdot\nabla)a = 0, \quad a(x,0) = x\\
&\partial_t q + (u\cdot\nabla)q + q[\mathbf{R}^*(C) : \nabla u + \mu\,\frac{h^2}{q^2}\psi_q^*(C)] = 0\\
&\partial_t h + (u\cdot\nabla)h + h[\mathbf{S}^*(C) : \nabla u + \mu\,\frac{h^2}{q^2}\psi_h^*(C)] = 0
\end{aligned} \tag{3}$$

with $G = \nabla a$, $C = G^T\, G$, $q = \frac{1}{2}\langle|w|^2\rangle$, $h = \langle w.\nabla_y \times w\rangle$.

Here, q and h are the turbulent kinetic energy and the turbulent helicity. The closure terms $\mathbf{R}^*$, $\mathbf{S}^*$ are 3×3 tensors and ψ_q^*, ψ_h^* are scalar functions. They depend only on ∇a, via the "universal" fluctuation $\tilde{w}^*$, that verifies a generalized Euler equation in the fast variable $\tau = t/\epsilon^{\frac{2}{3}}$, $y = a(x,t)/\epsilon$:

$$\begin{aligned}
&\partial_\tau\tilde{w}^* + (\tilde{w}^*\cdot\nabla_y)\tilde{w}^* + (C\nabla_y)\pi^* = 0, \quad \nabla_y\cdot\tilde{w}^* = 0\\
&\tilde{w}^*(y,0) = \frac{1}{\sqrt{q_0}}w^0(\frac{q_0}{h_0}y),\\
&\tilde{w}^*,\ \pi^* \qquad \text{almost-periodic in } (y-\tau).
\end{aligned} \tag{4}$$

Here, π^* is the pressure associated to $\tilde{w}^*$. Also, q_0 and h_0 are the mean kinetic energy and mean helicity of the initial perturbation w^0, respectively, assumed to be nonzero.

If w_0 is isotropic and h_0/q_0 is constant then this system of equations reduces to

$$\begin{aligned} &\partial_t u + (u.\nabla)u + \nabla p + \epsilon^{\frac{2}{3}}\nabla.[q^0(a(x,t))m(i)\nabla a \nabla a^T] = 0, \\ &\nabla.u = 0 \\ &\partial_t a + u\cdot\nabla a = 0, \qquad a(x,0) = x, \end{aligned} \tag{5}$$

with $i = \sum_{l,j} \partial_j a_l \partial_j a_l$, $\gamma_0 \approx \frac{1}{9}$, $\beta_0 \approx \frac{1}{3}$ and

$$m(i) = \frac{\beta_0}{(1+i)^2} e^{-\mu\gamma_0 t} e^{\frac{2\beta_0}{3}\frac{i-2}{1+i}}$$

3. CONNECTION WITH $k - \epsilon$

The full model above is a three equations model for the mean flow involving the turbulent kinetic energy q, the lagrangian coordinate a and the helicity h.

For a decaying homogeneous turbulence (u=0) equations (3) reduce to $k - \epsilon$. Indeed (3) becomes

$$\partial_t q + \frac{h^2}{q}\mu\psi_q^*(C) = 0,$$

$$\partial_t h + \frac{h^3}{q^2}\mu\psi_q^* h(C) = 0,$$

from these one can derive an equation for the turbulent rate of viscous dissipation $e = \mu\langle|\nabla \times w|^2\rangle$ because, by definition of ψ_q^* (see below), $e = \mu\frac{h^2}{q}\psi_q^*$. Therefore

$$\partial_t e + (2\frac{\psi_q^*}{\psi_h^*} - 1)\frac{e^2}{q} = 0$$

This is to be compared with the equation in IV.4.1, giving $2\frac{\psi_q^*}{\psi_h^*} - 1 \approx 2.06$.

The model has also some similarities with one equation models regarding energy dissipation. This is seen from the conservation of energy equation. If the equation for u is multiplied by u and integrated over R^3 then by adding the result to the equation for q, it is found that

$$\partial_t \int_{R^3} [\frac{u^2}{2} + \epsilon^{\frac{2}{3}} q] = -\mu\epsilon^{\frac{2}{3}} \int_{R^3} (\frac{h}{q})^2 \psi_q^*.$$

For many practical applications, h/q is constant (see below) so the turbulent diffusion is controlled only by the function $\psi_q^*(C, q)$, like in a one equation model where the rate of turbulent dissipation e is an algebraic function of q and ∇u.

The model does not support Reynolds' hypothesis because R^* is function of $\nabla a \nabla a^T$ instead of $\nabla u + \nabla u^T$. However it is important to notice that R^* is a transient term, much like a memory of the initial shape of the small structures.
Consider the case of a flat plate with a periodic initial distribution of eddies for w^0. Then the shear due to the mean flow deforms the periodic cubes into parallelograms and an analysis of $\nabla a \nabla a^T$ in this case shows that R^* acts like a recalling force so that the equation for u coupled with the one for a forms a wave equation when the deformations are small; Chacon[1988] observed similar oscillations in direct simulations.
As we shall see, the model is derived from an asymptotic expansion. If the second order term $u^{(2)}$ is kept in the result, then R has a dissipation term of order $\epsilon^{4/3}$. Let $\bar{u}^\epsilon = u + \epsilon^{2/3}\bar{u}^{(1)} + \epsilon\bar{u}^{(2)}$; then

$$\partial_t \bar{u}^\epsilon + (\bar{u}^\epsilon \cdot \nabla)\bar{u}^\epsilon + \nabla \bar{p}^\epsilon + \epsilon^{2/3}\nabla \cdot \tilde{\mathbf{R}}^* =$$
$$\epsilon^{4/3}\nabla \cdot [\nu_t(C)[\sqrt{q}(\nabla \bar{u}^\epsilon + (\nabla \bar{u}^\epsilon)^T)]],$$

where ν_t is an eddy viscosity fourth order tensor. It is theoretically possible to compute ν_t and its dependence upon C but the algebra is too complex.

In conclusion this multiple scales expansion shows that R does not depend only upon $\nabla u + \nabla u^T$ but also on $\nabla a \nabla a^T$.

4. CASCADE OF EQUATIONS

In the following $\langle u \rangle$ and $\langle\langle u \rangle\rangle$ will denote the space and space-time averages:

$$\langle u \rangle = \lim_{R \to \infty} \frac{1}{|B(x, R)|} \int_{B(x,R)} u(x, t, y, \tau) dy,$$

$$\langle\langle u\rangle\rangle = \lim_{T\to\infty}\frac{1}{T}\int_0^T \langle u\rangle(x,t,\tau)d\tau \tag{6}$$

We shall assume that u^ϵ, p^ϵ admit asymptotic expansions of the form

$$\begin{aligned}
&u^\epsilon(x,t) \sim u(x,t) + \epsilon^{1/3}w(\frac{a(x,t)}{\epsilon},\frac{t}{\epsilon^{2/3}};x,t)+\\
&\epsilon^{2/3}u^{(1)}(\frac{a(x,t)}{\epsilon},\frac{t}{\epsilon^{2/3}};x,t) + \epsilon u^{(2)}(\frac{a(x,t)}{\epsilon},\frac{t}{\epsilon^{2/3}};x,t)+\cdots\\
&p^\epsilon(x,t) \sim p(x,t) + \epsilon^{1/3}p^{(0)}(\frac{a(x,t)}{\epsilon},\frac{t}{\epsilon^{2/3}};x,t)+\\
&\epsilon^{2/3}\pi(\frac{a(x,t)}{\epsilon},\frac{t}{\epsilon^{2/3}};x,t) + \epsilon p^{(1)}(\frac{a(x,t)}{\epsilon},\frac{t}{\epsilon^{2/3}};x,t)+\cdots
\end{aligned} \tag{7}$$

Here $a(x,t)$ is the inverse Lagrangian coordinates associated to the velocity u:

$$\partial_t a + (u\cdot\nabla)a = 0, \quad \text{in } R^3\times]0,T[; \quad a(x,0) = x.$$

We shall assume that all functions w, $u^{(k)}$, π, $p^{(k)}$ are smooth and almost-periodic in the $y-\tau$ variables.
To obtain formal equations for the terms of expansions (7), at first one has to express equations (1) as expansions in powers of ϵ. In condensed form, these expansions are:

$$\begin{aligned}
&\epsilon^{-2/3}\{(\partial_t a + u\cdot\nabla a)\cdot\nabla_y\tilde{w} + C\nabla_y p^{(0)}\}+\\
&+\epsilon^{-1/3}\{(\partial_t a + u\cdot\nabla a)\cdot\nabla_y\tilde{u}^{(1)} + E(\tilde{w},\pi,C)\}+\\
&+\epsilon^{0}\{(\partial_t a + u\cdot\nabla a)\cdot\nabla_y\tilde{u}^{(2)}\\
&+[L(\tilde{u}^{(1)},p^{(1)};\tilde{w},C) - \tilde{f}^{(1)}]\}+\\
&+\epsilon^{1/3}\{(\partial_t a + u\cdot\nabla a)\cdot\nabla_y\tilde{u}^{(3)} + [L(\tilde{u}^{(2)},p^{(2)};\tilde{w},C) - \tilde{f}^{(2)}]\}+\\
&+\epsilon^{2/3}\{(\partial_t a + u\cdot\nabla a)\cdot\nabla_y\tilde{u}^{(4)} + [L(\tilde{u}^{(3)},p^{(3)};\tilde{w},C) - \tilde{f}^{(3)}]\}\\
&+\cdots = 0
\end{aligned}$$

at $y = \frac{a(x,t)}{\epsilon}$, $\tau = \frac{t}{\epsilon^{2/3}}$. Also,

$$\begin{aligned}
&\epsilon^{-2/3}(\nabla_y\cdot\tilde{w}) + \epsilon^{-1/3}(\nabla_y\cdot\tilde{u}^{(1)} - g^{(1)}) + \epsilon^{0}(\nabla_y\cdot\tilde{u}^{(2)} - g^{(2)})+\\
&+\epsilon^{1/3}(\nabla_y\cdot\tilde{u}^{(3)} - g^{(3)}) + \epsilon^{2/3}(\nabla_y\cdot\tilde{u}^{(4)} - g^{(4)}) + \cdots = 0.
\end{aligned}$$

at $y = \frac{a(x,t)}{\epsilon}$, $\tau = \frac{t}{\epsilon^{2/3}}$, with

$$C = G^TG, \quad \text{with} \quad G = \nabla a, \quad (\text{i.e.,} \quad G_{ij} = \frac{\partial a_j}{\partial x_i});$$

$$\tilde{w} = G^T w, \qquad \tilde{u}^{(k)} = G^T u^{(k)} \quad \forall\, k \geq 1;$$

$$E(\tilde{w}, \pi; C) = \partial_\tau \tilde{w} + (\tilde{w}\cdot\nabla_y)\tilde{w} + C\nabla_y \pi;$$

$$L(\tilde{u}^{(k)}, p^{(k)}; \tilde{w}, C) = \partial_\tau \tilde{u}^{(k)} + (\tilde{w}\cdot\nabla_y)\tilde{u}^{(k)} + (\tilde{u}^{(k)}\cdot\nabla_y)\tilde{w} + C\nabla_y p^{(k)}.$$

Also, functions $\tilde{f}^{(1)},\ \tilde{f}^{(2)},\ \cdots, g^{(1)}, g^{(2)},\ \cdots$ are defined by

$$\begin{aligned}
\tilde{f}^{(1)} &= -G^T[\partial_t u + (u\cdot\nabla)u + \nabla p] \\
\tilde{f}^{(2)} &= -G^T[\partial_t w + (u\cdot\nabla)w + (w\cdot\nabla)u + \nabla p^{(0)} \\
&\quad + (\tilde{u}^{(1)}\cdot\nabla_y)u^{(1)} - \mu\,(G\nabla_y)\cdot((G\nabla_y)w)] \\
\tilde{f}^{(3)} &= -G^T[\partial_t u^{(1)} + (u\cdot\nabla)u^{(1)} + (u^{(1)}\cdot\nabla)u + \nabla\pi \\
&\quad + (\tilde{u}^{(1)}\cdot\nabla_y)u^{(2)} + \\
&\quad + (\tilde{u}^{(2)}\cdot\nabla_y)u^{(1)} + (w\cdot\nabla)w - \mu\,(G\nabla_y)\cdot((G\nabla_y)u^{(1)})] \\
&\ \vdots
\end{aligned}$$

$$\begin{aligned}
g^{(1)} &= 0 \\
g^{(2)} &= -\nabla\cdot u \\
g^{(3)} &= -\nabla\cdot w \\
g^{(k)} &= \nabla\cdot u^{(k-3)}, \qquad \forall\, k \geq 4
\end{aligned}$$

Now we write that the sum is zero for all ϵ:

$$G\nabla_y p^{(0)} = 0,$$

$$E(\tilde{w}, \pi; C) = 0; \qquad \nabla_y\cdot\tilde{w} = 0; \tag{8}$$

$$L(\tilde{u}^{(k)}, p^{(k)}; \tilde{w}, C) = \tilde{f}^{(k)}, \qquad \nabla_y\cdot\tilde{u}^{(k)} = g^{(k)} \quad \forall k \geq 1. \tag{9}$$

5. STUDY OF $\tilde{w}$

Proposition
$\langle\tilde{w}\rangle$, $\frac{1}{2}\langle\tilde{w}\cdot C^{-1}\tilde{w})\rangle$ and $\langle\tilde{w}C^{-1}\tilde{r})\rangle$ are independent of τ because:

$$\partial_\tau\tilde{w} + \nabla_y\cdot[\tilde{w}\otimes\tilde{w} + C\pi] = 0, \tag{10}$$

$$\partial_\tau(\frac{1}{2}\tilde{w}\cdot C^{-1}\tilde{w}) + \nabla_y\cdot[\tilde{w}(\frac{1}{2}\tilde{w}\cdot C^{-1}\tilde{w} + \pi)] = 0, \tag{11}$$

$$\partial_\tau(\tilde{w}C^{-1}\tilde{r}) + \nabla_y\cdot[\tilde{w}(\tilde{w}\cdot C^{-1}\tilde{r}) + \tilde{r}(\pi - \frac{1}{2}\tilde{w}\cdot C^{-1}\tilde{w})] = 0, \tag{12}$$

where $\tilde{r} = \nabla_y \times (G^{-1}\tilde{w})$

Proof: (Sketch) Equations (10) and (11) are both direct consequences of (8). To obtain (12), let us follow Ortegon ([1989]): If $\det(G) = 1$, then

$$G^T[(G\nabla_y) \times \tilde{w}] = \nabla_y \times (C^{-1}\tilde{w}). \tag{13}$$

By taking the curl of (8), this formula allows to obtain an evolution equation for $\tilde{r}$:

$$\partial_\tau \tilde{r} + (\tilde{w} \cdot \nabla_y)\tilde{r} - (\tilde{r} \cdot \nabla_y)\tilde{w} = 0, .$$

Now, combining (8) and (13) yields (12).

Proposition
Assume the pair $(\tilde{w}^*, \pi^*)$ *is any smooth solution of Euler equations (8). Then, each pair* $(\tilde{w}, \pi)$ *of the five-parameters family given by*

$$\begin{aligned} \tilde{w}(y,\tau) &= \rho \tilde{w}^*(\lambda(y - \vec{\alpha}\tau), \rho\lambda\tau) + \vec{\alpha} \\ \pi(y,\tau) &= \rho^2 \pi^*(\lambda(y - \vec{\alpha}\tau), \rho\lambda\tau) \end{aligned} \qquad \forall \vec{\alpha} \in R^3, \quad \forall \lambda, \rho \in \mathbf{R} \tag{14}$$

is also solution of (8).

The proof is straightforward. Note that there is just one choice of parameters ρ, λ and $\vec{\alpha}$ in (14) that allows to match all τ-invariants of $\tilde{w}$:

Proposition
Assume that the initial mean kinetic energy and helicity:

$$q_0(x) = \frac{1}{2}\langle |w^0|^2 \rangle,$$

$$h_0 = h_0(x) = \frac{1}{2}\langle w^0 \cdot (\nabla_y \times w^0) \rangle,$$

are non zero.
Assume also that the pair $(\tilde{w}^*, \pi^*)$ *is a smooth solution of the generalized Euler problem (4) and that the curl of* $\tilde{w}^*$, $\tilde{r}^* = \nabla_y \times (C^{-1}\tilde{w}^*)$ *is almost-periodic. Then, for each* $\tilde{m} \in R^3$, $q \geq 0$, $h \in R$, *the velocity* $\tilde{w}$ *given by (14) with*

$$\rho = \sqrt{\tilde{q}}, \quad \lambda = \frac{h}{\tilde{q}}, \quad \vec{\alpha} = \tilde{m}, \quad \textit{with } \tilde{q} = q - \frac{1}{2}\tilde{m} \cdot C^{-1}\tilde{m},$$

verifies

$$\langle \tilde{w} \rangle = \tilde{m}, \quad \frac{1}{2}\langle \tilde{w} \cdot C^{-1}\tilde{w} \rangle = q, \quad \frac{1}{2}\langle \tilde{w} \cdot C^{-1}\tilde{r} \rangle = h. \tag{15}$$

Proof (Sketch): Note that the initial condition $\tilde{w}_0^*$ of problem (4) has zero mean and unit mean kinetic energy and mean helicity. Moreover, as $\tilde{w}^*$, $\tilde{r}^*$ are almost-periodic in y, the functions $\tilde{w}$, $\tilde{w} \cdot C^{-1}\tilde{w}$ and $\tilde{w} \cdot C^{-1}\tilde{r}$ are also almost-periodic. Then, the averages $\langle \tilde{w} \rangle$, $\langle \tilde{w} \cdot C^{-1}\tilde{w} \rangle$ and $\langle \tilde{w} \cdot C^{-1}\tilde{r} \rangle$ exist. Now, a straightforward calculation proves (15) (Cf. Ortegon [1989] for details).

In what follows, the analysis needs the working hypothesis that this problem admits a unique almost-periodic solution, which depends smoothly on y, τ and the parameter matrix C. This appears difficult to prove, as there are no results about existence of smooth global solutions of Euler equations in three space dimensions.

6. HIGHER ORDER TERMS

As the initial conditions for Navier-Stokes (2) do not contain terms in ϵ of order superior to 1/3, an appropriate initial condition for (9) is

$$\tilde{u}^{(k)} = 0 \text{ at } t = 0$$

This initial condition avoids a given solution $(\tilde{u}^{(k)}, p^{(k)})$ of (9) to have a family of associated solutions similar to that given in Proposition 2 for Euler equations (8). Indeed, if $(\tilde{u}_p, q_p)$ is a solution of (9), then the general solution will be of the form

$$\tilde{u}^{(k)} = \tilde{u}_p + \tilde{u}_h; \qquad p^{(k)} = q_p + q_h,$$

where $(\tilde{u}_h, q_h)$ is the general solution of the homogeneous problem:

$$\begin{aligned} &L(\tilde{u}_h, q_h; \tilde{w}, C) = 0, \qquad \nabla_y \cdot \tilde{u}_h = 0 \\ &\tilde{u}_h(x, 0) = 0 \end{aligned}$$

This equation does not possess as many symmetries as Euler equations (8). The only solutions of (9) similar to (14) are given by

$$\begin{aligned} &\tilde{u}_h(y,\tau) = \rho \tilde{u}_h^*(y - \vec{\alpha}\tau, \lambda\tau); \\ &\tilde{w}(y,\tau) = \lambda \tilde{w}^*(y - \vec{\alpha}\tau, \lambda\tau) + \vec{\alpha}; \qquad \forall\, \vec{\alpha} \in R^3, \quad \forall \rho, \lambda \in R; \\ &q_h(y,\tau) = \rho\lambda q_h^*(y - \vec{\alpha}\tau, \lambda\tau) \end{aligned}$$

where $(\tilde{u}_h^*, q_h^*)$ is solution of (9) and $\tilde{w}^*$ is a solution of (8). It is now clear that the only solution that matches the initial conditions of (9) is $\tilde{u}_h \equiv 0$.

Proposition

The following equations hold:

$$\begin{aligned}&\partial_\tau \tilde{u}^{(k)} + \nabla_y \cdot [\tilde{w} \otimes \tilde{u}^{(k)} + \tilde{u}^{(k)} \otimes \tilde{w} + Cp^{(k)}] \\ &= \tilde{f}^{(k)} + \tilde{w} g^{(k)},\end{aligned} \tag{16}$$

$$\begin{aligned}&\partial_\tau(\tilde{w} \cdot C^{-1}\tilde{u}^{(k)}) + \nabla_y \cdot [\tilde{w}(\tilde{w} \cdot C^{-1}\tilde{u}^{(k)} + p^{(k)}) \\ &+ \tilde{u}^{(k)}(\frac{1}{2}\tilde{w} \cdot C^{-1}\tilde{w} + \pi)] = \\ &= \tilde{w} \cdot C^{-1}\tilde{f}^{(k)} + (\frac{1}{2}\tilde{w} \cdot C^{-1}\tilde{w} + \pi) g^{(k)},\end{aligned} \tag{17}$$

$$\begin{aligned}&\partial_\tau(\tilde{w} \cdot C^{-1}\tilde{s}^{(k)} + \tilde{u}^{(k)} \cdot C^{-1}\tilde{r}) \\ &+ \nabla_y \cdot [\tilde{w}(\tilde{w} \cdot C^{-1}\tilde{s}^{(k)} + \tilde{u}^{(k)} \cdot C^{-1}\tilde{r}) \\ &+ \tilde{s}^{(k)}(\pi - \frac{1}{2}\tilde{w} \cdot C^{-1}\tilde{w})] + \\ &+ \nabla_y \cdot [\tilde{r}(p^{(k)} - \tilde{w} \cdot C^{-1}\tilde{u}^{(k)}) + \tilde{u}^{(k)}(\tilde{w} \cdot C^{-1}\tilde{r})] \\ &= \tilde{w} \cdot C^{-1}\tilde{F}^{(k)} + \tilde{r} \cdot C^{-1}\tilde{f}^{(k)},\end{aligned} \tag{18}$$

where $\tilde{r} = \nabla_y \times (C^{-1}\tilde{w})$, $\tilde{s}^{(k)} = \nabla_y \times (C^{-1}\tilde{u}^{(k)})$ *and* $\tilde{F}^{(k)} = \nabla_y \times (C^{-1}\tilde{f}^{(k)})$.

Proof (Sketch)

Equations (16) and (17) are direct consequences of (8) and (9). Also, taking the curl of (9) gives a transport equation for $\tilde{s}^{(k)}$:

$$\partial_\tau \tilde{s}^{(k)} + (\tilde{w} \cdot \nabla_y)\tilde{s}^{(k)} +$$

$$(\tilde{u}^{(k)} \cdot \nabla_y)\tilde{r} - (\tilde{s}^{(k)} \cdot \nabla_y)\tilde{w} - (\tilde{r} \cdot \nabla_y)\tilde{u}^{(k)}$$

$$= \tilde{F}^{(k)} - \tilde{r} g^{(k)}$$

Combining this equations with (8), (9) and the definition of $\tilde{r}$ yields (17) (Cf.[13]).

Proposition

i) Assume that $\tilde{w}$, π, $\tilde{u}^{(k)}$ *and* $p^{(k)}$ *are almost-periodic and smooth enough. Then the following compatibility conditions hold:*

$$\ll \tilde{f}^{(k)} + \tilde{w}g^{(k)} \gg = 0, \tag{19}$$

$$\ll \tilde{w}\cdot C^{-1}\tilde{f}^{(k)} + (\frac{1}{2}(\tilde{w}\cdot C^{-1}\tilde{w}) + \pi)g^{(k)} \gg = 0, \tag{20}$$

$$\langle g^{(k)}(\cdot,\tau)\rangle = 0, \qquad \forall\, \tau \in \mathbf{R} \tag{21}$$

ii) Assume in addition that $\tilde{r}$ and $\tilde{s}^{(k)}$ are almost-periodic. Then, the following compatibility condition also holds:

$$\ll \tilde{r}\cdot C^{-1}\tilde{f}^{(k)} \gg = 0.$$

Proof (Sketch):
To prove (19), for instance, let us average equation (16) in $B_R \times [-\tau,\tau]$. This yields:

$$\frac{1}{2\tau\,|B_R|}\Big[\int_{B_R}\Big(\tilde{u}^{(k)}(y,\tau)dy - \tilde{u}^{(k)}(y,-\tau)\Big)\,dy +$$

$$+\int_{-\tau}^{\tau}\int_{\Gamma_R}\vec{n}(y,\sigma)\cdot(\tilde{w}\otimes\tilde{u}^{(k)} + \tilde{u}^{(k)}\otimes\tilde{w} + Cp^{(k)})(y,\sigma)d\Gamma_R(y)\,d\sigma\Big] =$$

$$= \frac{1}{2\tau\,|B_R|}\int_{-\tau}^{\tau}\int_{B_R}(\tilde{f}^{(k)} + \tilde{w}g^{(k)})(y,\sigma)dy\,d\sigma$$

As $\tilde{w}$, $\tilde{u}^{(k)}$ and $p^{(k)}$ are bounded, the l.h.s vanishes as R and τ tend to infinity. Thus, (19) follows. The proof of (20) and (21) are similar. Also, averaging the last equation in (9) on B_R yields

$$\frac{1}{|B_R|}\int_{\Gamma_R}(\vec{n}\cdot\tilde{u}^{(k)})(y,\tau)d\Gamma_R(y) = \frac{1}{|B_R|}\int_{B_R}g^{(k)}(y,\tau)dy$$

As $\tilde{u}^{(k)}$ is bounded, the l.h.s vanishes as $R\uparrow\infty$. Thus, (21) follows.

In what follow we shall assume that (9) admits a unique smooth almost-periodic solution $(\tilde{u}^{(k)}, p^{(k)})$ when the compatibility conditions (19)-(21) are fulfilled.

7. AVERAGED EQUATIONS

Averaged equations are obtained in a natural way by implementing the compatibility conditions (19)-(21) for $k = 1, 2, \cdots$. In the case $k = 1$ we obtain Euler equations for the pair (u, p) :

$$\partial_t u + (u \cdot \nabla)u + \nabla p = 0, \quad \nabla \cdot u = 0.$$

When $k = 2$, transport equations for the invariants of w are obtained:

1) **Mean** $m = \langle w \rangle$:

$$\partial_t m + (m \cdot \nabla)u + (u \cdot \nabla)m + \nabla \bar{p}^{(0)} = 0; \quad \nabla \cdot m = 0,$$

with $\bar{p}^{(0)} = \ll p^{(0)} \gg$.

2) **Mean Kinetic Energy** q :

$$\partial_t q + (u \cdot \nabla)q + \mathbf{R} : \nabla u + \mu \psi_q = 0,$$

where the tensor $\mathbf{R}$ and the scalar function ψ_q are closure terms, defined by

$$\mathbf{R} = \ll w \otimes w \gg, \qquad \psi_q = \ll \tilde{r} \cdot C^{-1} \tilde{r} \gg .$$

3) **Mean Helicity** h :

$$\partial_t h + (u \cdot \nabla)h + \mathbf{S} : \nabla u + \mu \psi_h = 0,$$

where the closure terms $\mathbf{S}$ and ψ_h are defined by

$$\mathbf{S} = \ll w \otimes r \gg, \qquad \psi_h = \ll \tilde{r} \cdot C^{-1}(\nabla_y \times (C^{-1}\tilde{r})) \gg .$$

When $k = 3$, a linear Reynolds equation for the mean velocity field

$$\bar{u}^{(1)} = \ll u^{(1)} \gg$$

appears:

$$\partial_t \bar{u}^{(1)} + (u \cdot \nabla)\bar{u}^{(1)} + (\bar{u}^{(1)} \cdot \nabla)u + \nabla \bar{p}^{(1)} + \nabla \cdot \mathbf{R} = 0, \quad \nabla \cdot \bar{u}^{(1)} = 0,$$

with $\bar{p}^{(1)} = \ll p^{(1)} \gg$.
Therefore

$$m \equiv 0, \qquad \bar{p}^{(0)} \equiv \text{constant}.$$

Now, it is possible to give explicit formulae for the closure terms , as a consequence of the symmetries of the microstructure problem (9). A careful calculation of closure terms leads to (Cf.[13]):

$$\mathbf{R}(q, h; G) = q\, G^{-T} \tilde{\mathbf{R}}^*(C) G^{-1}; \quad \psi_q(q, h; G) = \frac{h^2}{q}\, \psi_q^*(C);$$

$$\mathbf{S}(q, h; G) = h\, G^{-T} \tilde{\mathbf{S}}^*(C) G^{-1}; \quad \psi_h(q, h; G) = \frac{h^3}{q^2}\, \psi_h^*(C);$$

where

$$\tilde{\mathbf{R}}^*(C) = \ll \tilde{w}^*(C) \otimes \tilde{w}^*(C) \gg;$$
$$\psi_q^*(C) = \ll \tilde{r}^*(C) \cdot C^{-1}\tilde{r}^*(C) \gg;$$
$$\tilde{\mathbf{S}}^*(C) = \ll \tilde{r}^*(C) \otimes \tilde{w}^*(C) \gg;$$
$$\psi_h^*(C) = \ll \tilde{r}^*(C) \cdot C^{-1}(\nabla_y \times C^{-1}\tilde{r}^*(C)) \gg .$$

Now, the system of averaged equations, including the Lagrangian coordinates, is the following:

$$\partial_t u + (u \cdot \nabla)u + \nabla p = 0, \quad \nabla \cdot u = 0 \quad u(x,0) = u_0(x)$$
$$\partial_t a + (u \cdot \nabla)a = 0, \quad a(x,0) = x$$
$$\partial_t q + (u \cdot \nabla)q + q \left[(G^{-T}\tilde{\mathbf{R}}^*(C)G^{-1}) : \nabla u + \mu \left(\frac{h}{q}\right)^2 \psi_q^*(C) \right] = 0$$
$$q(x,0) = q_0(x) = \frac{1}{2}\langle |w^0(\cdot,x)|^2 \rangle$$
$$\partial_t h + (u \cdot \nabla)h + h \left[(G^{-T}\tilde{\mathbf{S}}^*(C)G^{-1}) : \nabla u + \mu \left(\frac{h}{q}\right)^2 \psi_h^*(C) \right] = 0$$
$$h(x,0) = h_0(x) = \frac{1}{2}\langle w^0(\cdot,x) \cdot (\nabla_y \times w^0)(\cdot,x) \rangle$$
$$\partial_t \bar{u}^{(1)} + (u \cdot \nabla)\bar{u}^{(1)} + (\bar{u}^{(1)} \cdot \nabla)u$$
$$+ \nabla \bar{p}^{(1)} + \nabla \cdot (qG^{-T}\tilde{\mathbf{R}}^*(C)G^{-1}) = 0$$
$$\nabla \cdot \bar{u}^{(1)} = 0 \quad u^{(1)}(x,0) = 0$$

Solving this system, we obtain an approximation to the average value of u^ϵ of (theoretical) order ϵ, given by $u + \epsilon^{2/3}\bar{u}^{(1)}$.

8. LOCAL ISOTROPY

Still $\tilde{R}^*(C), \psi_q^*(C), \psi_h^*(C)$ are not known. By assuming local isotropy it becomes possible to apply rotational invariance and derive formulae for these closure terms.

Definition 1
Let $v : R^3 \to R$ be a vector function. We shall say that v is isotropic if $v(Qy) = Qv(y)$, for all $y \in R^3$, for all rotation matrix Q.

Definition 2
Let $v(C;\cdot) : R^3 \to R^3$ be an almost-periodic vector function for each 3×3 matrix C.

We shall say that v is statistically isotropic at matrix C if

$$\langle v_{i_1}(Q^TCQ;\cdot),\cdots,v_{i_k}(Q^TCQ;\cdot)\rangle$$
$$= Q_{j_1i_1}\cdots Q_{j_ki_k}\langle v_{j_1}(Q^TCQ;\cdot),\cdots,v_{j_k}(Q^TCQ;\cdot)\rangle,$$

for all integer $k \geq 1$, for all $i_1,\cdots,i_k \in 1,2,3$ and for all rotation matrix Q.

We shall say that $\tilde{w}^*$, solution of (8), is statistically isotropic if it is statistically isotropic at any 3×3 symmetric matrix C with determinant one.

Let us now analyze the dependence of the canonical fluctuation $\tilde{w}^*$ upon the matrix C:

Proposition

If the initial perturbation w^0 is isotropic, then, for each rotation matrix Q we have

$$\begin{aligned}
&\tilde{w}^*(Q^TCQ;y,\tau) = Q^T\tilde{w}^*(C;Qy,\tau)\\
&\pi^*(Q^TCQ;y,\tau) = \pi^*(C;Qy,\tau)\\
&\quad\forall\, y\in R^3, \quad \forall\, \tau \in R.
\end{aligned} \tag{22}$$

Proof: (Sketch)

Let us define the following transformations:

$$Y = Q^Ty, \quad \tilde{W}^*(Y,\tau) = Q^T\tilde{w}^*(C;y,\tau);$$

$$\Pi^*(Y,\tau) = \pi^*(C;y,\tau).$$

Then, the pair $(\tilde{W}^*,\Pi^*)$ verifies:

$$\begin{aligned}
&\partial_\tau\tilde{W}^* + (\tilde{W}^*\cdot\nabla_Y)\tilde{W}^* + (Q^TCQ)\nabla_Y\Pi^* = 0, \quad \nabla_Y\cdot\tilde{W}^* = 0,\\
&\tilde{W}^*(Y,0) = Q^T\tilde{w}_0^*(Qy)\\
&\tilde{W}^*,\ \Pi^* \qquad \text{almost-periodic in } (y-\tau).
\end{aligned}$$

As w^0 is isotropic, then $\tilde{w}_0^*$ is also isotropic and $(\tilde{W}^*,\Pi^*)$ is a solution of (8) with matrix Q^TCQ instead of C. Under the hypothesis of unicity of solutions of (8), this yields (22)

Corollary

Under the hypotheses of Proposition 6, the fluctuation $\tilde{w}^*$ is statistically isotropic. In particular, the canonical closure terms verify

$$\begin{aligned}\tilde{\mathbf{R}}^*(Q^T C Q) = Q^T \tilde{\mathbf{R}}^*(C) Q, \quad \psi_q^*(Q^T C Q) = \psi_q^*(C),\\ \tilde{\mathbf{S}}^*(Q^T C Q) = Q^T \tilde{\mathbf{S}}^*(C) Q, \quad \psi_h^*(Q^T C Q) = \psi_h^*(C),\end{aligned} \tag{23}$$

for all rotation matrix Q, for all 3×3 symmetric matrix C with determinant one.

Proof: The statistical isotropy of $\tilde{w}^*$ is an immediate consequence of (22) and the invariance under rotations of the space average operator defined in (11). Also, (23) follows directly from (22) for the same reasons.

These structures can be used now to simplify notably the dependence of closure terms on the matrix C as follows:

Proposition

Let G be a 3×3 matrix with determinant one and $\bar{C} = GG^T$. Let us denote by $i_1 = trace(\bar{C})$ and $i_2 = trace(Adj(\bar{C}))$ the two nontrivial invariants of $\bar{C}$. Then, under the hypotheses of Proposition 6, the following statements hold:

i) There exist six functions α_0, α_1, α_2, β_0, β_1, β_2, which depend only upon i_1 and i_2, such that

$$\begin{aligned}\mathbf{R}(q, h; G) = q[\alpha_0(i_1, i_2) I + \alpha_1(i_1, i_2)\bar{C} + \alpha_2(i_1, i_2)\bar{C}^2];\\ \mathbf{S}(q, h; G) = h[\beta_0(i_1, i_2) I + \beta_1(i_1, i_2)\bar{C} + \beta_2(i_1, i_2)\bar{C}^2].\end{aligned}$$

ii) There exist two functions $\tilde{\psi}_q$, $\tilde{\psi}_h$, which depend only upon i_1 and i_2, such that

$$\psi_q(q, h; G) = \frac{h^2}{q}\tilde{\psi}_q(i_1, i_2); \quad \psi_h(q, h; G) = \frac{h^3}{q^2}\tilde{\psi}_h(i_1, i_2).$$

Proof: i) Due to (23), tensor $\tilde{\mathbf{R}}^*$ verifies the hypotheses of Rivlin-Ericksen's theorem (see chapter 3). Then, there must exist three functions $\tilde{\alpha}_0$, $\tilde{\alpha}_1$ and $\tilde{\alpha}_2$ depending only upon the two nontrivial invariants $\tilde{i}_1$ and $\tilde{i}_2$ of C, such that

$$\tilde{\mathbf{R}}^*(C) = \tilde{\alpha}_0(\tilde{i}_1, \tilde{i}_2) I + \tilde{\alpha}_1(\tilde{i}_1, \tilde{i}_2) C + \tilde{\alpha}_2(\tilde{i}_1, \tilde{i}_2) C^2.$$

Let us recall now that matrix C verifies its characteristic polynomial (Cayley-Hamilton's Theorem):

$$C^3 - \mathrm{trace}(C) C^2 + \mathrm{trace}(\mathrm{Adj}(C)) C - \det(C) = 0.$$

Moreover, trace(C) $= i_1$ and trace(Adj(C)) $=$ trace(C^{-1}) $=$ trace($\bar{C}^{-1}$) $= i_2$. Then, the first conclusion of i) follows, with

$$\alpha_0 = \tilde{\alpha}_1 + i_2\tilde{\alpha}_0; \quad \alpha_1 = \tilde{\alpha}_2 - i_1\tilde{\alpha}_0; \quad \alpha_2 = \tilde{\alpha}_0$$

A similar deduction holds for tensor $\tilde{S}^*$.

ii) From (23) it follows that ψ_q^* and ψ_h^* must depend only upon the invariants of C. Then, invariance and dimensional analysis yields the conclusion of ii).

Proposition

Assume the canonic fluctuation $\tilde{w}^*$ is locally statistically isotropic. Then model equations (3) are frame-invariant.

9. REDUCTION OF THE MODEL

In dimension 2 the Reynolds tensor is such that:

$$\nabla \cdot [\tilde{\mathbf{R}}^*(C)] = \nabla \cdot [\tilde{\alpha}_1(\tilde{i}_1)C]$$

Furthermore if h/q is constant at initial time then frame invariance implies that the equation for h is the same as the one for q and so h/q remains constant at all time.

Numerical simulation of w shows that

$$\tilde{\alpha}_1(\tilde{i}_1) \sim q\beta_0 \frac{\tilde{i}_1}{(1+\tilde{i}_1)^2}$$

Therefore model (5) is obtained.

9.1 A $k - \epsilon$ model with memory

In (2) the stress tensor is not viscous in the sense that it conserves energy when $\mu = 0$:

$$\int_\Omega (\frac{1}{2}u^2 + q^0(a(x,t)e^{\frac{2\beta_0}{(n+1)} - \frac{2\beta_0}{(1+i)}}) = constant$$

To account for dissipation one must pursue further the asymptotic expansion that lead to (2) ; then a term proportional to $\sqrt{q}(\nabla u + \nabla u^T)$ is found and the model becomes

$$\partial_t u + (u.\nabla)u + \nabla p + \nabla.[\epsilon^{\frac{2}{3}} \frac{\beta_0 q}{(1+i)^2} \nabla a \nabla a^T - \epsilon \nu_0 \sqrt{q}(\nabla u + \nabla u^T)] = 0,$$

in R^n, where

$$q = q^0(a(x,t))e^{-\mu\gamma_0 t(x)} e^{\frac{2\beta_0}{(n+1)} - \frac{2\beta_0}{(1+i)}}$$

with $\gamma_o = 1/9$. But it was found in Begue et al[36] that this is too crude a model for the viscous effect of the turbulence. Since the $k-\epsilon$ model is quite good to simulate the effect of dissipation, it is possible to built a mixed model where the kinetic turbulent energy is the product of the one coming from the model above and the one coming from the $k-\epsilon$ model and similarly. The rate of dissipated turbulent energy is the product of the kinetic energy q and of ϵ found in the $k-\epsilon$ model. The Reynolds tensor is the sum of the tensors of both models (q is given above) :

$$\begin{aligned}
&R = k\beta_0 \frac{e^{\frac{2\beta_0(i-2)}{3(1+i)}}}{(1+i)^2} \nabla a \nabla a^T - c_\mu \frac{k^2}{\epsilon}(\nabla u + \nabla u^T) \\
&\partial_t k + u\nabla k - \frac{c_\mu}{2}\frac{k^2}{\epsilon}|\nabla u + \nabla u^T|^2 - \nabla.(c_\mu \frac{k^2}{\epsilon}\nabla k) + \epsilon = 0 \\
&\partial_t \epsilon + u\nabla\epsilon - \frac{c_1}{2}k|\nabla u + \nabla u^T|^2 - \nabla.(c_\epsilon \frac{k^2}{\epsilon}\nabla\epsilon) + c_2\frac{\epsilon^2}{k} = 0 \\
&i = \sum_{l,j} a_{l,j}^2 \\
&\partial_t a + u\nabla a = 0, \quad a(x,0) = x
\end{aligned} \tag{24}$$

Notice that by changing the value of the constant β_0 and q^0, we can pass continuously from the $k-\epsilon$ model to the model obtained by homogenization.

10. NUMERICAL EXPERIMENTS

11.1 Solution of (4) for $\tilde{w}^*$

Problem (4) is a Euler equation with periodic conditions and a parameter C in the pressure. All the closure terms of the turbulence model can be computed by solving (4) for various values of the tensor C.
This task was attempted by Begue[1983] and pursued by Saiac[1990] and Chacon[1988]. In Begue, the equations are discretized by a finite

element method and solved by least square by a conjugate gradient method. In Saiac and Chacon, the equations are solved in time over a sufficiently long time interval, by a finite element scheme which preserves the invariants. Although it is difficult to obtain a precise answer, estimates for ψ_q^*, ψ_h^* where obtained and (5) was verified to be reasonnable numerically.

10.2 Comparison with a direct simulation

Once the coefficients were known, the model was compared with a direct simulation. For this a Fourrier-Chebichev based computer program was given to by Patera[1984] for the simulation of channel flows between two parallel flat plates and with periodic conditions between the inflow and outflow boundaries (figure 10.1).

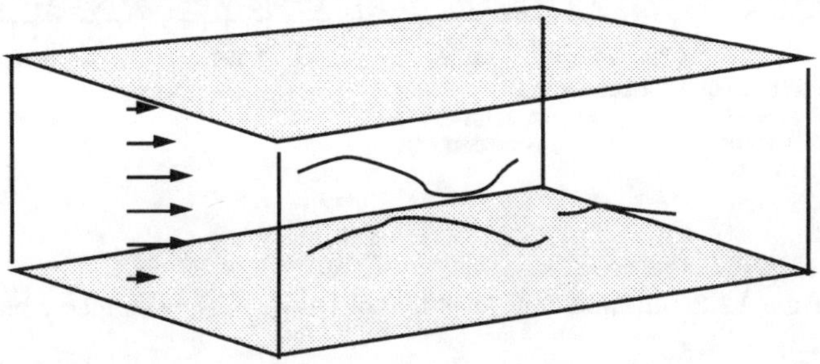

Figure 12.1 Turbulent flow between two flat plates: the geometry

Problem (1)(2) was simulated directly with $N \times N \times M$ modes, with w_0 having modes in the second half of the spectrum only: $|k| > N/2$. Accordingly u_o has modes only in the first half and a spectral gap was kept between u_0 and w_0.

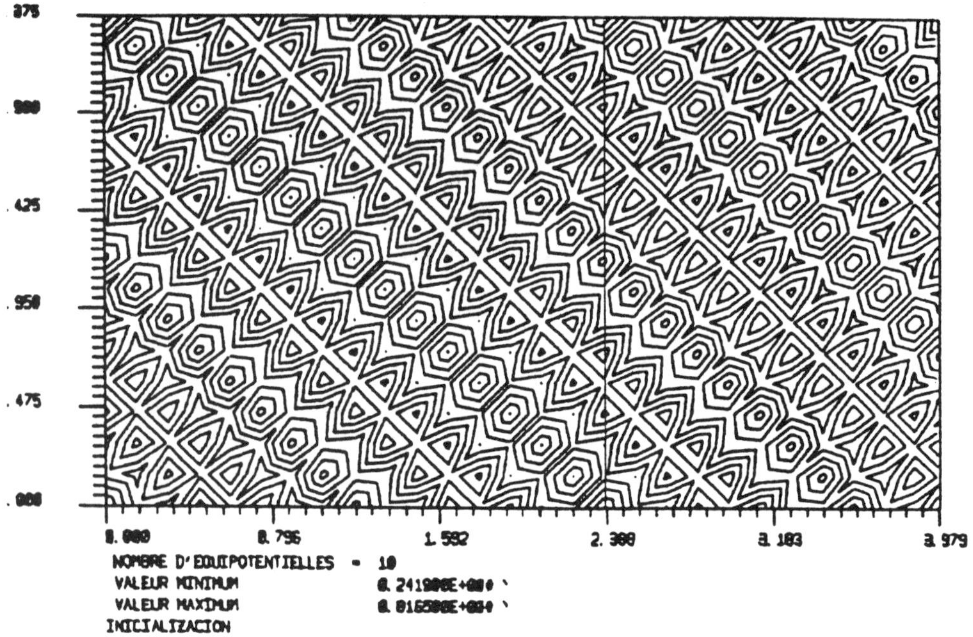

Figure 12.2 Initialization: level line of $||u_0||$ in the symmetry plane

It was found that q is globally decreasing in time (viscous effects) but with oscillations. In other words, $||\log q||_0$ is an oscillatory function of time (figure 12.3).

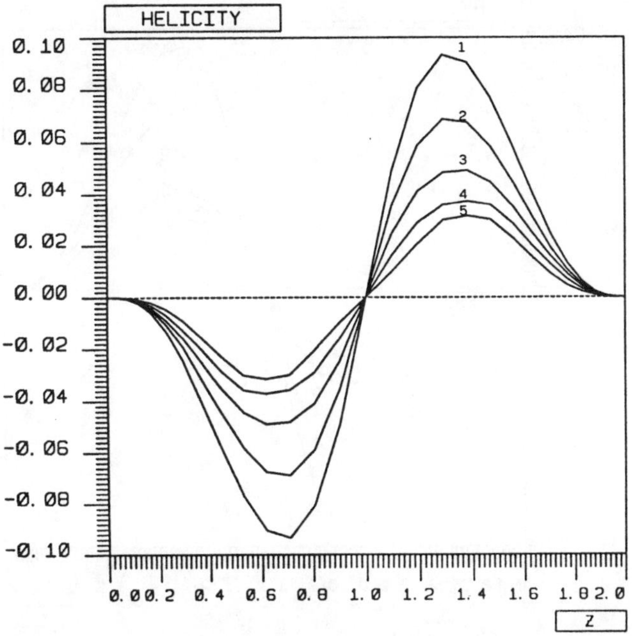

Figure 12.3 Turbulent flow between two flat plates: direct simulation

The computation is done with $N = 16$ and $M = 32$. This figure shows the rate of decay of the kinetic turbulent energy as a function of time for $\epsilon = 1/5$ and $\epsilon = 1/10$.

Then these results were compared with the model, which in this case, is one dimensional in space. for small q_0 equations (5) is hyperbolic in character and the frequency of the oscillations can be found. They where found to be correctly predicted. Thus this numerical experiment shows that the high modes can cause the low modes to oscillate in time.

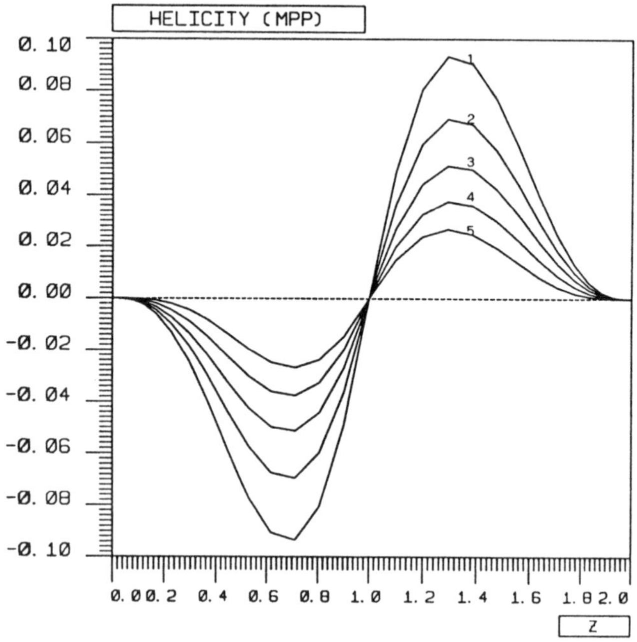

Figure 12.4 Comparison between the model and the direct simulation

These figures show q (left) and h (right) in a plane perpendicular to the mean flow as a function of x_3 and t. The to two figures are obtained with the direct simulation while the bottom ones are obtained with the model.

10.3 Comparison with the $k - \epsilon$ model

In Cardot[1989] a comparison with the $k - \epsilon$ model is made for a two dimensional flow over a cylinder. It uses model (24) and the numerical method described in Chapter 4 for the same cylinder. There it was found that the term proportional to $\nabla a \nabla a^T$ is on the average, 10% in size, that of the term porportional to $\nabla u + \nabla u^T$. So its contribution os not critical and it may be neglected[1]

It is not clear however that the situation does not change when the grid size (i.e. ϵ) is decreased. Thus it may be that this analysis is a direction for future research aimed at establishing a $k - \epsilon$ "subgrid scale" model where the eddy viscosity would be a vanishing function of the mesh size.

1 This may explain why these oscillatory behaviors, predicted by the mathematics, have been altogether neglected or even undetected by engineers.

REFERENCES

Achenbach E. : Distribution of Local Pressure and Skin-Friction Around a Cicrular Cylinder in a Cross-Flow up to $Re = 5.10^6$, *J. Fluid Mech.,* Vol. 34, pp. 625-639, 1968.

AFOSR-HTTM: Stanford conference on complex turbulent flows, 1980-1981.

Aliabadi S., T. Tezduyar: Space-Time Finite Element Computation of Compressible Flows Involving Moving Boundaries and Interfaces, *AHPCRC report,* 1993.

Anderson J.D., Jr.: Modern compressible flow with historical perspective. McGraw Hill. 1982. see also *Fundamental Aerodynamics* Mc-Graw Hill, 1984.

Arnold D., F. Brezzi, M. Fortin: A stable finite element for the Stokes equations. *Calcolo* 21(4) pp. 337-344, 1984.

Avelaneda M., A. Majda: Mathematical Models with Exact Renormalization for Turbulent Transport. *Comm. Math. Phys.* , 1992.

Babuska I., W. Rheinboldt: Error estimates for adaptive finite element computations. *SIAM J. Numer Comp.* Vol. 15, 1978.

Bachelor G.K. : An Introduction to Fluid Dynamics. *Cambridge University Press,*1970.

Baldwin B., T. Barth: A one equation turbulent transport model for high Reynolds wall bounded flows. *NASA* TM-102847, 1991.

Baldwin B., H. Lomax: Thin layer approximation and algebraic model for separated turbulent flows. *AIAA-paper* 78-257, Huntsville, 1978.

Baker A. : Finite element computational fluid mechanics. McGraw-Hill, 1985.

Beale T.J., A. Majda: Rate of convergence for viscous splitting of the Navier-Stokes equations. *Math. Comp.* Vol. 37, pp. 243-260, 1981.

Begue C. : Simulation de la turbulence par méthode d'homogénéisation. Thèse, Université Paris 6, 1983.

Begue C., B. Cardot, C. Parès, O. Pironneau: Simulation of turbulence with transient mean *Int. J. Num. Meth. in Fluids,* Vol. 11, 1990.

Benque J.P., B. Ibler, A. Keraimsi, G. Labadie: A finite element method for the Navier-Stokes eq. *Norries ed.* , pp. 1110-1120, Pentech Press, 1980.

Benqué J.P., O. Daubert, J. Goussebaile, H. Haugel: Splitting up techniques for computations of industrial flows. In *Vistas in applied mathematics.* A.V. Balakrishnan ed. Optimization Software inc. Springer, 1986.

Bensoussan A., J.L. Lions, G. Papanicolaou: Homogenisation methods for partial differential equations. Dunod 1982.

Bergé P., Y. Pomeau, Ch. Vidal: De l'ordre dans le chaos. Hermann, 1984.

Bernardi C., G. Raugel: A conforming finite element method *SIAM J. Numer Anal,* Vol. 22, Num. 3, 1985.

Boris J., D.L. Book: Flux corrected transport. *J. Comp.. Phys.* Vol. 20, pp. 397-431, 1976.

Brachet M.: To appear in J. Fluid Mech

Brezzi,F.: On the existence, uniqueness and approximation of saddle-point problems arising from Lagrange multipliers. *RAIRO Anal Numer.* Vol. 8, pp. 129-151, 1974.

Brun G. : Développements et application d'une méthode d'élements finis pour le calcul d'écoulements compressibles turbulents fortement chauffés, Thèse Ecole Centrale de Lyon, 1988.

Cafarelli L., R. Kohn, L. Nirenberg: On the regularity of the solutions of the Navier-Stokes equations. *Comm. Pure and Appl. Math.*, Vol. 35, pp. 771-831, 1982.

Cardot B.: Simulation d'ecoulement turbulent par un modele k-epsilon. These de doctorat, Universite Paris 6. 1989.

Carrau A.: Modelisation numerique d'un ecoulement sur paroi rugeuse. Thèse de Doctorat, 1992.

Cebeci T., A. Smith: Analysis of turbulent boundary layers, *Academic Press,* 1974.

Chacon T. : Etude d'un modele pour la convection des microstructures. These, Université Paris 6, 1985.

Chacon T.: Oscillations due to the transport of micro-structures. *SIAM J. Appl. Math.* Vol. 48, Num.5, pp. 1128-1146,1988.

Chacon T., O. Pironneau: On the mathematical foundations of the $k - \epsilon$ model. In *Vistas in applied mathematics.* A.V. Balakrishnan ed. Optimization Software inc. Springer, 1986.

Chacon T., D Franco, F. Ortegon: Helicity in turbulence modelling by homogenization. To appear in Comp. Fluid Dyn. 1993.

Chalot F., Z. Johan, M. Mallet, M. Ravachol, G. Rogé: Development of a finite elmeent Navier-Stokes Solver with application to turbulent and hypersonic flows. *AIAA paper* 92- 0670, 1992.

Childress S, A. Soward: Scalar transport and alpha effect for a family of cat's eye flows. *J. Fluid Mech.* Vol. 205, pp. 99-133, 1989.

Chollet J.P.: Turbulence tridimensionelle isotrope. *Thèse de doctorat,* Grenoble, IMG, 1984.

Chollet J.P., M. Lesieur: Parametrization of small scales of 3D isotropic turbulence using spectral closures. *ICTAM meeting, Toronto August,* 1980.

Cebeci T., A. Smith: Analysis of turbulent boundary layers, *Academic Press,* 1974.

Chorin A., J. Marsden: An introduction to Fluid Mechanics, *Springer Universitext,* 1979.

Ciarlet Ph. : The Finite Element Method for Elliptic problems. *North-Holland,* 1978.

Ciarlet Ph.: Elasticité tridimensionnelle, *Masson Ed,* 1986.

Coakley T.: Numerical simulation of viscous transonic airfoil flows. *AIAA paper,* 87-0416. 1987.

Colella P., P. Woodward: The Piecewise Parabolic Method (PPM) for gas dynamic simulations.*J. Comp. Phys,* Vol. 174, 1984.

Comte-Bellot G. : Ecoulement turbulent entre deux parois planes.*Publications scientifiques et techniques du ministère de l'air,* 1980.

Comte-Bellot G. : Turbulence. *Lecture notes. Ecole Centrale de Lyon,* 1980.

Conca, C: Etude d'un fluide traversant une paroi preforée. *J. Math Pures et appl.*, Vol. 66, pp. 45-69, 1987.

P. Constantin, C. Fioas, O. Manley, R. Temam: Determining modes and fractal dimension of turbulent flows.*J. Fluid Mech.* Vol.150, pp. 427-440, 1985.

Cousteix J. : Couche limite laminaires et turbulentes, Tomes 1 and 2. *Cepadues-editions*, 1990.

David E.: Modélsiation des écoulements compressibles et hypersoniques: une approche instationnaire. *Thèse*, INPG, Grenoble, 1993.

Deardorff J.W. : A numerical study of 3-d turbulent channel flow at large Reynolds numbers. *J. Fluid Mech.* Vol. 41, Num. 2, pp. 453-480, 1970.

Delery J., M.C. Coet: Experiments on shock-wave/boundary-layer interactions produced by two-dimensional ramps and three-dimensional obstacles, *Proc. of the Workshop on Hypersonic Flows for Reentry Problems,* Vol. 2, Springer Verlag, Antibes 22-26 jan. 1990.

Denman P., J. Harvey, R. Hillier: Hypersonic Boundary Layer and Base Flow. *Proc. of the Workshop on Hypersonic Flows for Reentry Problems,* Vol. 2, Springer Verlag, Antibes 22-26 jan. 1990.

Dervieux A. : Steady Euler simulations using unstructured meshes. *Von Karman Lecture notes*, series 1884-04. Mars, 1985.

Desideri J.A., A. Dervieux: Compressible Fluid Dynamics; compressible flow solvers using unstructured grids. *Von Karman lecture notes series.* March, 1988.

Diperna R. : Convergence of approximate solutions to conservation laws. *Arch Rat. Mech. Anal.* Vol. 82, pp. 27, 1983.

diPerna R., A. Majda: The validity of Nonlinear Geometric Optics for weak solutions of Conservation laws. *Comm. Math. Phys.* Vol. 98, pp. 313-347, 1985.

Douglas J., T.F. Russell: Numerical methods for convection dominated diffusion problems based on combining the method of characteristics with finite element methods or finite difference method. *SIAM J. Numer Anal.* Vol. 19, Num. 5, pp. 871-885, 1982.

E W.: Propagation of oscillations in the solutions of 1-d compressible fluid equations. *Courant Institute Report,* 1990.

Edgar B., Woodward P.: Diffraction of a shock wave by a wedge. *AIAA report,* 1991.

A. Fannjiang, G. Papanicolaou: Convection enhanced diffusion for periodic flows. *submitted to SIAM J. Applied Math.* 1993.

Favre A., L.S.G.Kovasnay, R. Dumas, J. Gaviglio, M. Coantic: La turbulence en mécanique des fluides. *Ed. Gauthier Villars*, 1976.

Feder J: *Fractals.* Plenum Press. 1988.

Fezoui L: Resolution des équations d'Euler par un schéma de Van Leer en éléments finis. *Rapport Inria* 358, 1985.

Foias P., O. Manley, R. Temam: Determing modes and Fractal dimension of turbulent flows. *J. Fluid Mech*, Vol. 150, pp. 427-440, 1985.

Ghidaglia, J.M: On the fractal dimension of attractors for viscous incompressible flows. *SIAM J. Numer. Anal,* Vol. 17, pp. 1139-1157, 1986.

Girault V., P.A. Raviart: Finite Elements methods for Navier-Stokes equations. *Springer series SCM*, Vol. 5, 1986

Glowinski R. : Numerical Methods for Nonlinear Variational Methods. *Springer Series in Computational physics*, 1984.

Godlewski E., P.A. Raviart: Systemes hyperboliques. *Collection SMAI: math et application.* Ellipse, 1992.

Goldberg U. : Separated flow treatment with a new turbulence model, *AIAA Journal* Vol.24, Num. 10, pp. 1711-1713. 1986.

Gorski J. : A new near-wall formulation for the $k - \varepsilon$ equations of turbulence, *AIAA paper* Num.86-0556, 1986.

Goussebaïle J., A. Jacomy: Application à la thermo-hydrolique des méthodes d'éclatement d'opérateur dans le cadre éléments finis: traitement du modèle $k - \varepsilon$. *Rapport EDF-LNH* HE/41/85.11,1985.

Grasso F., C. Speziale: Supersonic flow computations by two-equation turbulence modelling. *AIAA paper* 89-1951, 1989.

Guillard H.: Simulation of turbulent flows with unstructured meshes. (To appear). 1994.

Haase W., F. Brandsma, E. Elsholz, M. Leschziner, D. Schwamborn: EUROVAL- An European Initiative on Validation of CFD Codes. *Notes on numerical fluid dynamics,* vol 42. Vieweg, 1993.

Hanjalic K., B. E. Launder: Contribution towards a Reynolds-stress closure for low-Reynolds number turbulence, *J. Fluid Mech.,* Vol. 74, pp. 593-610, 1976.

Harris C.: Two-Dimensional Aerodynamic Characteristics of the NACA 0012 in the Langley 8-Foot Transonic Pressure Tunnel, *NASA report* TM 81927, 1981.

Hasholder J.M., M. Mallet, A. Naim, P. Rostand: Problemes aerothermiques locaux pour la rentree d'Hermes. *AAAF workshop 29,* Biscarosse. 1992.

Hecht F., C. Pares: NSP1B3 un logiciel pour resoudre les equations de Navier-Stokes incompressible. *Rapport de recherche INRIA*, RR1449. 1991.

HERMES: Proc. of the Workshop on Hypersonic Flows for Reentry Problems, Vol. 1 and 2, Springer Verlag, Antibes 22-26 jan. 1990.

Hinze J.: Turbulence. McGraw-Hill, 1975.

Hirsch, C.: Numerical computation of internal and external flows. Wiley, 1990.

Holts T. : Viscous Transonic Airfoil Workshop Compendium of Results, *AIAA paper*, 87-1460.

Hou T., X. Xin: Homogenization of linear transport equations. with oscillatory vector fields. *SIAM J. Appl. Math.*, Vol. 52, Num.1, pp. 34-45, 1992.

Hughes T. J.R. : The finite element method. *Prentice Hall,* 1987.

Hughes T. J.R., T. Franca, M. Mallet: A New Finite Element Formulation for Computational Fluid Dynamics: I. Symmetric Forms of the Compressible Euler and Navier-Stokes Equations and the Second Law of Thermodynamics. *Computer Methods in Applied Mechanics and Engineering*, Num. 54, pp. 223-234, 1986.

Hughes T.J.R., M. Mallet: A New Finite Element Formulation for Computational Fluid Dynamics: III. The Generalized Streamline Operator for Multidimensional

Advective-Diffusive Systems. *Computer Methods in Applied Mechanics and Engineering,* Num. 58, pp. 305-328, 1986.

Hughes T.J.R., M. Mallet: A new finite element formulation for computational fluid dynamics. *Computer Meth. in Appl. Mech. and Eng.* Num. 54, pp. 341-355, 1986.

Hughes T.J.R., M. Mallet: A New Finite Element Formulation for Computational Fluid Dynamics: IV. Discontinuity Capturing Operator for Multi-Dimensional Advective-Diffusive Systems. *Computer Methods in Applied Mechanics and Engineering,* Num. 58, pp. 329-339, 1986.

Hughes T.J.R., T. Tezduyar: Finite Element Methods for First-Order Hyperbolic Systems with particular Emphasis on the Compressible Euler Equations. *Computer Methods in Applied Mechanics and Engineering,* Num. 45, pp. 217-284, 1984.

Hur N., S. Thangam, C. Speziale: Numerical study of turbulent secondary flows in curved ducts. *ICASE report*, 89-25, 1989.

Hutton A.G., R.M. Smith, S. Hickmott: The computation of turbulent flows of industrial complexity by the finite element method. Progress and prospects. *Finite éléments in fluids.* Vol. 7, Num. 15, 1987.

Hytopoulos K., R. Simpson: Critical evaluation of recent second-order closure models. *AIAA paper* 93-0081, 1993.

Ikeda T.: Maximum principle in finite element method for convection-diffusion pheneomena, *Lect. note in num. and appl. analysis,* Vol. 4, Num. 76. North-Holland/Kinikuniya. 1976

Isichenko M.: Percolation, Statistical topography and transport in random media. *Rev Mod. Phys.* Vol. 64, pp. 961, 1992.

Jaeger M. : Simulation d'écoulements turbulents par elements finis tridimensionels. *Thèse*, Université de Compiegne, 1990.

Jansen K., Z. Johan, T.J.R. Hughes: Implementation of a one equation turbulence model within a stabilzed FEM formulation of a symmetric advective-diffusive system. *25th INRIA anniversary,* Proc. Bensoussan ed. to appear

Jaubertau F., C. Rosier, R. Temam: The nonlinear Galerkin method. *Appl. Numer. Mech.* 1990.

Johnson C. : Numerical solution of PDE by the finite element method. *Cambridge university press*, 1987.

Johnson C. : Streamline diffusion methods for problems in fluid mechanics. *Finite elements in fluids* , Vol 6. R. Gallagher ed. Wiley, 1987.

Johnson C., A. Szepessy: On the convergence of streamline diffusion finite element methods for hyperbolic conservation laws. *Numerical methods for compressible flows*, T.E. Tedzuyar ed. AMD-78.

Kandula M., P. Buning: Evaluation of Baldwin-Barth turbulence model with thin layer Navier-Stokes computation of an axisymmetric afterbody-exhaust jet flowfield. *AIAA paper*, 93-0418, 1993.

Kesten H., G. Papanicolaou: A limit theorem for stochastic acceleration. *Commun. Math. phys.* Vol. 78, pp. 19-63, 1980.

Kolmogorov A. : The equations of turbulent motion in an incompressible fluid. *Isv. Acad. Sci. USSR Phys.* Vol. 6, pp. 56-58, 1942.

Kraichnan R. EDQNM, *Phys. Fluids.* Vol. 7, pp. 1163, 1964.

Ladyzhenskaya O. : The Mathematical Theory of viscous incompressible flows. *Gordon-Breach,* 1963.

Lam C., A. Bremhorst: Modified form of the $k - \varepsilon$ model for predicting wall turbulence. *Journal of Fluids Ingineering,* Vol. 103, p 456-460, 1981.

Landau L., E. Lifschitz: Mecanique des Fluides, *MIR*, Moscou, 1953.

Launder B.E.: On the modelling of turbulent industrial flows. *Computer Methods in Applied Sciences* Ch. Hirsch ed. Elsevier, 1992.

Launder B.E., D.B. Spalding: Mathematical models of turbulence. *Academic press,* 1972.

Launder B.E., Reece G.J., Rodi W.: Progress in the development of a Reynolds stress turbulence closure, *J.F.M,* Vol. 68, Part 3, 1975.

Laurence D.: Modelling in industrial aerodynamics, *EDF-DER report* HE-412/90.13A, 1990.

Lawrence S. : Hypersonic Cone Flow Prediction Using an Implicit Upwind Space-Marching Code, *Proc. of the Workshop on Hypersonic Flows for Reentry Problems,* Vol. 2, Springer Verlag, Antibes 22-26 jan. 1990.

Lax P, S. Burstein, A. Lax: Calculus with application and computing. *Spring Int. Student ed.* 1979.

Lebeau G., T. Tezduyar: Finite Element Computation of Compressible Flows with the SUPG Formulation. *Advances in Finite Element Analysis in Fluid Dynamics*, Eds. M.N. Dhaubhadel, M.S. Engelman and J.N. Reddy, FED-Vol.123, ASME, New York, 1991.

Lesieur M.: Turbulence in fluids. *Martinus Nijhoff publishers,* 1987.

Lewandowski R., B. Mohammadi: Existence and positivity results for the $\phi - \theta$ model and a modified $k - \epsilon$ turbulence model. *Math model and methods in applied sicences,* Vol. 3, Number 2, pp 195-215, 1993.

Lewandowski R: Modlèles de turbulences et équations paraboliques. to appear as Note CRAS, 1993.

Lions J.L.: Quelques Methodes de Resolution des Problèmes aux limites Nonlineaires. Dunod, 1968.

Lions P.L.: Existence of solution for compressible isentropic viscous flows. *Seminaire Univ Paris 6,* (to appear).

Lumley J.L.: Prediction methods for turbulent flows, *VKI Lectures* 76, 1975.

Mallet M. : A finite element method for computational fluid dynamics. *Ph.D. thesis,* University of Stanford, 1985.

Martinelli L., Yakhot V.: RNG-Based turbulence transport approximations with application to transonic flows. *AIAA report* 89-1950-CP, 1989.

Maslov V: Homogenization of Navier-Stokes equations. *Seminaire du Collège de France,* H. Brezis, J.L. Lione ed. Pitmann, 1987.

Matsumura A., T. Nishida: The initial value problem for the equations of motion of viscous and heat conductive gases. *J. Math. Kyoto Univ.* Vol. 20, pp. 67-104, 1980.

Matsumura A., T. Nishida: Initial boundary value problems for the equations of motion in general fluids, *Computer Methods in applied sciences and Engineering.* R. Glowinski et al. eds; North-Holland, 1982.

Mclaughin D., G. Papanicolaou, O. Pironneau: Convection of microstructures. *Siam J. Numer. Anal.* Vol. 45, Num. 5, pp. 780-796, 1982.

Metais O., M. Lesieur: Spectral large eddy simulation of isotropic and stably-stratified turbulence. *J. Fluid Mech.*, Vol. 239, pp. 157-194, 1992.

Mohammadi B.: A stable Finite Element algorithm for the $k-\varepsilon$ model for compressible turbulence, *INRIA report*, 1991.

Mohammadi B.: Complex Turbulent Flows Computation with a Two-Layer Approach,*Int. J. Num. Meth. Fluids*, Vol. 15, pp. 747-771, 1992.

Mohammadi B.: Finite element compuation of compressible turbulent flows with the kepsilon model,*Revue Europeenne des Elements Finis,* Vol.2, Number 1, pp 26-46, 1993.

Mohammadi B., O. Pironneau: Analysis of the $k - \epsilon$ model. J. Wiley, 1993.

Mohammadi B., J.H. Saiac: Turbulent Compressible Axisymmetric flows Computation with the $k - \varepsilon$ Model, *Computational Fluid Dynamics*, Vol 1, pp. 1125-133, 1993.

Moin P., J. Kim: Large eddy simulation of turbulent channel flow. *J. Fluid. Mech.* Vol.118, pp. 341, 1982.

Murat F: Compacité par compensation. Ann. Sc. Norm. Sup. Pisa, 5, p 489-507, 1978.

Netterfield M.: Computation of Hypersonic Turbulent Flow over Rearward facing Step. *Proc. of the Workshop on Hypersonic Flows for Reentry Problems,* Vol. 2, Springer Verlag, Antibes 22-26 jan. 1990.

Neveu J.: Mathematical foundations of the calculus of Probability, *Holden Day,* 1965.

Nguetseng G.: A general convergence result for a functional related to the theory of homogenization. *SIAM J. Appl. Math.*, Vol. 20, pp. 608-623, 1989.

Orszag S.: Statistical theory of turbulence. 1973.

Orszag S., A. Patera: Secondary instability of wall bounded shear flows. *J. Fluid Mech.* Vol. 128, pp. 347-385, 1983.

Ortegon F. : Modelisation des écoulement turbulents à deux échelles par méthodes d'homogénéisation. These, Université Paris 6, 1989.

Osher S.: Convergence of generalized MUSCL schemes. *ICASE report*, 172-306, 1984.

Panton R.L.: Incompressible flow. *Wiley Interscience,* 1984.

Papanicolaou G., O. Pironneau: On the asymptotic behaviour of motion in random flow. In Stochastic Nonlinear Systems, *Arnold-Lefever eds.*, Springer, 1981.

Parès C.: Un traitement faible par élément finis de la condition de glissement sur une paroi pour les équations de Navier-Stokes. *Note C.R.A.S.*, Vol. 307, pp.101-106, 1988.

Patankar, S.V.: Numerical heat transfer and fluid flow, *McGraw-Hill,* 1980.

Patel V.C., W. Rodi, G. Scheuerer: Turbulence models for near-wall and low Reynolds number flows : a review, *AIAA Journal*, Vol. 23, Num. 9. 1989

Patel V.C., H.C. Chen: Near-wall turbulence models for complex flows including separation, *AIAA Journal*, Vol. 29, Num. 6. 1989.

Perrier P., O. Pironneau: Subgrid turbulence modeling by homogenization. *Applied Math. Modeling,* Vol. 2, pp. 295-317. 1981.

Peyret R., T. Taylor: Computational methods for fluid flows. *Springer series in computational physics,* 1985.

Pironneau O.: On the transport-diffusion algorithm and its applications to the Navier-Stokes equations. *Numer. Math.* Vol. 38, pp. 309-332, 1982.

O. Pironneau: Finite Element Methods for Fluids, *Wiley,* 1989.

Porter D., A. Pouquet, P. Woodward: Kolmogorov-like spectra in decaying three-dimensional supersonic flow. *Physics of Fluids A.* (to appear), 1994.

Porter D., A. Pouquet, P. Woodward: A numerical study of supersonic turbulence. *Theoret. Comput. Fluid Dynamics*, Vol. 4, pp.13-49, 1992.

Prandtl L.: Uber die ausgebildete turbulenz. *ZAMM,* Vol. 5, pp. 136- 139, 1925.

Reynolds O.: On the dynamic theory of incompressible viscous fluids and the determination of the criterion. *Philos. Trans. R. Soc. London Ser. A* Vol.186, pp. 123-164, 1895.

Reynolds W. : Computation of turbulent flows. *Ann. Rev. Fluid Mech.* Vol. 8, pp. 183-208, 1975.

Ritchmeyer R., K. Morton: Difference methods for initial value problems. *Wiley,* 1967.

Rogallo R., P. Moin: Numerical simulation of turbulent flows. *Annual Rev. Fluid Mech.* Vol.16, pp. 99-137, 1984.

Rodi W. : Turbulence models and their application in hydraulics. Int. Ass. for hydraulic research, state-of-the-art *paper Delft,* 1980.

Rodi W.: The prediction of free boundary layers by use of a two-equation model of turbulence, *Ph.D. thesis*, University of London, 1972.

Rodi W.: Turbulence modeling for incompressible flows. *Euromech report*, 180. PHC Journal 7, 5, pp. 297-324, 1986.

Rose H., P.L. Sulem: Fully developed turbulence and statistical mechanics. *Journal de Physique,* Vol. 39, Num. 5, pp. 441-484, 1978.

Rosenhead L. ed.: Laminar Boundary layers. *Oxford University Press*, 1963.

Rostand Ph. : Sur une méthode de volumes finis en maillage non structuré pour le calcul d'écoulements visqueux compressibles.*Ph. D. Thesis,* Université Paris VI, 1989.

Rotta J. : Statistische Theorie nichthomogener turbulenz. *Z Phys.*, Vol. 129, pp.547-572, 1951.

Ruelle D: Les attracteurs étranges. La recherche, Vol. 108, pp.132, 1980.

Ruelle D: Characteristic exponents for a viscous fluid subjected to time dependent forces. *Com. Math. Phys*, Vol. 93, pp. 285-300,1984.

Ruelle D, F. Takens: On the nature of turbulence. *Commun. Math. Phys.*, Vol. 20, pp. 167-192, 1971.

Saad Y. : Krylov subspace methods for solving unsymmetric linear systems. *Math of Comp.* Vol. 37, pp. 105-126,1981.

Saiac J.H.: A time stepping scheme and a finite element method invariant preserving for the Euler equations. (To appear). 1994.

Schlichting H.: Boundary layer theory, Mc Graw-Hill, 1979.

Schumann V. : Realizability of Reynolds stress turbulence models. *Phys. Fluids.,* Vol. 20, pp. 721-725, 1977.

Schumann V. : subgrid scale model for finite difference simulations of turbulent flows in plane channel and annuli. *J. Comp. Physics.*, Vol. 18, pp. 376-404, 1975.

Serre D.: Quelques methodes d'étude de la propagation d'oscillations hyperboliques nonlinéaires. *Seminaire no 20 "Equations aux dérivées partielles,* Ecole Polytechnique Mars, 1991.

Serre D. : Oscillations nonlineaires des systemes hyperboliques. *Ann. Inst. Henri Poincaré.* , Vol. 8, Num. 3, pp. 351-417, 1991.

Shakib F. : Finite Element Analysis of the Compressible Euler and Navier Stokes Equations, *Ph.D. Thesis,* Stanford Univ. 1988.

Sinai G.: Introduction to Ergodic Theory. *Princeton University Press*, 1976.

Smagorinsky J.S. : General circulation model of the atmosphere. *Mon. Weather Rev.* Vol. 91, pp. 99-164,1963.

Spalard P., S. Allmaras: A one equation turbulence model for aerodynamic flows. *AIAA paper,* 92-0439, 1992.

Speziale C.: On non-linear K-l and $K-\epsilon$ models of turbulence. *J. Fluid Mech.* Vol. 178, pp 459-475,1987.

Speziale C.:Turbulence modeling in non-inertial frames of references. *ICASE report* 88-18,1988.

Speziale C., N. Mac Giolla Mhuiris: On the prediction of equilibrium states in homogenous turbulence,*J. Fluid Mech.*, vol. 209, pp. 591-615, 1989.

Stanisic, M.: The Mathematical Theory of Turbulence. *Springer Universitext,*1985.

Steger, J., R.Warming: Flux vector splitting of the invscid gas dynamics equations with applications to finite difference methods. *J. Comp. Physics* 40, pp 263-293, 1981.

Stoufflet B., J. Periaux, L. Fezoui, A. Dervieux: Numerical simulations of 3D hypersonic Euler flows around space vehicles using adaptive finite elements *AIAA* paper 8705660, 1987.

Strang G.: Fractional step methods. *SIAM J. Numer. Anal.* pp. 506, 968.

Tartar L.: Nonlocal effects induced by homogenization. In *PDE and Calcul of Variation* F. Culumbini ed. Birkhauser, 1989.

Tartar L. : Compensated compactness and applications. *Proc Nonlinear Analysis and Mechanics symposium,* Pittman RNM 39, 1979.

Tartar L. : H-measures, a new approach for studying homogenization, oscillation and concentration effects in PDE.*Proc Royal Soc Edimburg,* A. 115A, pp 193-230, 1990.

Thomasset F.: Implementation of Finite Element Methods for Navier-Stokes eq. *Springer series in Comp. Physics,*1981.

Valli A.: An existence theorem for compressible viscous flow. *Boll Un. Mat. It. Anal. Funz. Appl.* 18-C, pp317-325, 1981.

Vandromme D.: Contribution à la modélisation et la prédiction d'écoulements turbulents à masse volumique variable. *Thesis,* University of Lille, 1983.

Vandromme D., H. Haminh: Solution of the compressible Navier-Stokes equations with application to complex turbulent flows. *VKI-CFD course*, Rhodes St Génèse, Belgium, April 1984.

Viollet P.L.: On the modeling of turbulent heat and mass transfers for computations of buyoancy affected flows. *Proc. Int. Conf. Num. Meth. for Laminar and Turbulent Flows* . Venezia, 1981.

von Karman T.: Progress in the statistical theory of turbulence. *Proc. Natl. Acad. Sci. USA.* **34**, pp.530-539, 1948.

Weinan E: Homogenization of linear and nonlinear transport equations. *NYU Preprint,* 1991.

Woodward P.: Numerical techniques for gas dynamical problems in astrophysics. Computational Astrophysics, *P. Woodward Ed. Academic Press,* 1991.

Woodward P.: Scientific visualization of super computer simulations of fluid flow. (to appear).

Yakhot V., S. Orszag: Renormalization group analysis of turbulence, *Journal of Scientific Computing,***1**, 1986.

Yakhot V., S. Thangam, T. Gatski, S. Orszag, C.Speziale: Development of turbulence models for shear flows by a double expansion technique, *ICASE report* num. 91-65. july 1991.

Young L: Lectures on the calculus of variations and optimal control, *W.B. Saunders,* Philadelphia, 1969.

Zaidman S.: Almost periodic functions in abstract spaces, *Research notes in Math.* **126**, Pitman, 1985.

APPENDIX

NSC2KE : A Software for Turbulent Flows

NSC2KE is a 2D and axisymmetric finite volume / finite element Fortran solver for compressible Euler and Navier-Stokes equations in conservation form. The $k - \varepsilon$ turbulence model is implemented with various boundary conditions. This solver is based on techniques developped along this book and it can be considered as a support for its understanding.
The solver has been tested on several computers (e.g. APOLLO 425-DN10000, HP 735, IBM RS6000, CRAY-2, CRAY-C90).

Numerical techniques

The solver is splitted in two parts. The first part treats the viscous fluxes (if any) with a P1 finite element technique and the second part computes convective fluxes using finite volume. For the later part, the user can choose between Roe, Osher or Kinetic based approximate Riemann solvers. Two-layer and wall-law techniques are available for low-Reynolds number regions. The solver is second order in time and explicit and it runs on unstructured triangular meshes.
To give an idea of the CPU time, the complete configuration with turbulence takes around 0.001 seconds per node per iteration on an 1Mflops computer.

DATA Inputs

The user defines his computational configuration using the file **fort.1** which contains the following informations.

iaxi	=1 axisymmetric, =0 2D
ivis	=0 Euler, =1 Navier-Stokes
reynolds	Reynolds number per meter
xmach	inflow Mach number
twall	wall temperature in Kelvin
tinf	inflow temperature in Kelvin
iturb	=0 no turbulence model, =1 $k - \varepsilon$
ilaw	=0 no wall-law, =1 two-layer
delta	if(ilaw=1) δ for wall-law else the maximum width of the low-Reynolds region around the body
cfl	C.F.L. number ≤ 1

ktmax	maximum number of time step
iloc	=0 global time stepping, =1 local time stepping
ncont	=0 start with uniform mean flow variables, =1 restart from file **fort.20**
ncont1	=0 start with uniform turbulent variables, =1 restart from file **fort.21**
iflux	=1 roe, =2 osher, =3 kinetic

MESH Input

The program reads the mesh in routine **mailla.f** from **fort.14**. The requested format is:

read(14) number of points, number of triangles
read(14) ((coor(1,i),coor(2,i),logic(i)),i=1,number of points)
read(14) (nu(1,i),nu(2,i),nu(3,i)),i=1,number of triangles)

where coor(2,*) contains the coordinates of the points, nu(3,*) the connectivity table and logic(*) the type of boundary conditions to use.

Output and Results

The results are saved in routine **result.f**. The mean flow quantities are written in the following format on file **fort.10**:
read(10) (($\rho(i)$, $u(i)$, $v(i)$, $T(i)$),i=1,number of points)

The turbulent quantities are presented in the following format on file **fort.11**:
read(11) (($k(i)$, $\varepsilon(i)$, $\mu_{lam}(i)$, $\mu_{turb}(i)$),i=1,number of points)

where ρ is the density, u and v the velocity components, T the temperature, k the kinetic turbulent energy, ε the rate of turbulent dissipation, $\mu_{laminar}$ the laminar viscosity computed by the Sutherland law's and $\mu_{turbulent}$ the turbulent viscosity.
Of course, the user can easily change all the input and output procedures to make them more convenient for him.

How to get NSC2KE

You can obtain the NSC2KE by simple Email request at
mohamadi@menusin.inria.fr.
The automatic mesh generator EMC2 written by F.Hecht and E.Saltel may be obtained by the same channel.
All the examples in this book using unstructured meshes can be computed with NSC2KE.

INDEX

N, O, P, Q

R, S, T, U

V, W, X, Y, Z

MASSON Éditeur
120, boulevard Saint-Germain
75280 Paris Cedex 06
Dépôt légal : janvier 1994

Société des nouvelles éditions liégeoises, SA
Rue Saint-Vincent 12 - 4020 Liège
décembre 1993